About Island Press

Since 1984, the nonprofit Island Press has been stimulating, shaping, and communicating the ideas that are essential for solving environmental problems worldwide. With more than 800 titles in print and some 40 new releases each year, we are the nation's leading publisher on environmental issues. We identify innovative thinkers and emerging trends in the environmental field. We work with world-renowned experts and authors to develop cross-disciplinary solutions to environmental challenges.

Island Press designs and implements coordinated book publication campaigns in order to communicate our critical messages in print, in person, and online using the latest technologies, programs, and the media. Our goal: to reach targeted audiences—scientists, policymakers, environmental advocates, the media, and concerned citizens—who can and will take action to protect the plants and animals that enrich our world, the ecosystems we need to survive, the water we drink, and the air we breathe.

Island Press gratefully acknowledges the support of its work by the Agua Fund, Inc., Annenberg Foundation, The Christensen Fund, The Nathan Cummings Foundation, The Geraldine R. Dodge Foundation, Doris Duke Charitable Foundation, The Educational Foundation of America, Betsy and Jesse Fink Foundation, The William and Flora Hewlett Foundation, The Kendeda Fund, The Forrest and Frances Lattner Foundation, The Andrew W. Mellon Foundation, The Curtis and Edith Munson Foundation, Oak Foundation, The Overbrook Foundation, the David and Lucile Packard Foundation, The Summit Fund of Washington, Trust for Architectural Easements, Wallace Global Fund, The Winslow Foundation, and other generous donors.

The opinions expressed in this book are those of the author(s) and do not necessarily reflect the views of our donors.

OLD GROWTH IN A NEW WORLD

Old Growth in a New World

A PACIFIC NORTHWEST ICON REEXAMINED

EDITED BY

Thomas A. Spies and Sally L. Duncan

ISLANDPRESS

Washington • Covelo • London

Island Press is a trademark of The Center for Resource Economics.
No copyright claim is made in the works of Peter A. Bisson, Andrew B. Carey,
Eric D. Forsman, Gordon H. Reeves, Thomas A. Spies, and Fredrick J. Swanson,
employees of the federal government.

Chapter 15, "In the Shadow of the Cedars: The Spiritual Values of Old-Growth
Forests," by Kathleen Dean Moore, was previously published in *Conservation Biology* 21,
no. 4: 1120–23. Reprinted with the permission of Wiley-Blackwell.

Old growth in a New World : a Pacific Northwest icon reexamined /
edited by Thomas A. Spies and Sally L. Duncan.
p. cm.
Includes bibliographical references and index.
ISBN-13: 978-1-59726-409-9 (cloth : alk. paper)
ISBN-10: 1-59726-409-1 (cloth : alk. paper)
ISBN-13: 978-1-59726-410-5 (pbk. : alk. paper)
ISBN-10: 1-59726-410-5 (pbk. : alk. paper)
1. Old growth forests—Northwest, Pacific. 2. Old growth forest conservation—
Northwest, Pacific. 3. Old growth forest ecology—Northwest, Pacific.
I. Spies, Thomas A. (Thomas Allen) II. Duncan, Sally L.
SD387.O43O545 2008
333.75'1609795—dc22   2008008245

Printed on recycled, acid-free paper ✛

Manufactured in the United States of America

10 9 8 7 6 5 4 3 2 1

The saga of the relationship of Americans with their forests is by no means over. It is a continuous process to which there is no end.

M. WILLIAMS, *Americans and Their Forests*

It is a common feature of paradigm shifts . . . that they are dominated at first by generalizations, simplifications, and theory that is only correct some of the time and only in some situations. This has certainly been true for the old-growth paradigm shift. As the shortcomings of such generalizations and oversimplifications become apparent, there is normally an elaboration of the paradigm to incorporate the true complexity of the issue that is involved. It is time for this to happen in the case of old-growth forests.

J. P. KIMMINS, *Forestry Chronicle*

CONTENTS

ACKNOWLEDGMENTS

We would like to thank the National Commission on Science for Sustainable Forestry, especially John Gordon and Chris Bernabo, for supporting a workshop in the Columbia Gorge in Washington State in May 2005, which led to the idea for this book, and for assistance with the production of the book. We also thank the participants at that workshop for generating many of the ideas found in these pages. Jamie Barbour and the Focused Science Delivery Program of the U.S. Department of Agriculture–Forest Service, Pacific Northwest Research Station, provided funding to help with the development of the book.

We appreciate the reviews of the chapters that we received from many individuals. We especially thank John Cissel for his ideas and reviews of several of the chapters. Norm Johnson, Fred Swanson, and Denise Lach provided many helpful suggestions and ideas for the book during its development. Keith Olsen provided support for the web page we used to communicate during the writing of the book and designed several of the figures. Tami Lowry formatted and edited the book, Kathryn Ronnenberg reformatted or redesigned many of the figures, and Rob Pabst assisted with preparation of the final manuscript. Barbara Dean and Erin Johnson provided many good suggestions and saw this book through to publication.

PART I

Introduction and History

It is not possible to write a book about old-growth forests without considering ecological, social, and economic history. Old growth is, after all, about time and about how forests change from both natural and human causes. Thus, old growth also reflects *ideas* about forests, which co-evolve with scientific and cultural forces. In this opening part, we set the stage for reexamining the old-growth icon in the Pacific Northwest. The first chapter by Spies and Duncan introduces our search for better understanding in the 1980s as scientists began to study old-growth forests and as managers and the public began to debate their future. The period from 1980 to 1993 was characterized by a creative scientific and social ferment that turned management of federal forests upside down in the Pacific Northwest. The burden of proof shifted from those who would protect old growth to those who wanted to cut it. In the fifteen-year period since the adoption of old growth conservation policies on public lands, we've had an opportunity to reflect on how this change happened, examine how well the new approaches have worked, and investigate new ways of understanding this coupled natural–human system through ecological and social sciences. It is clear that the old-growth icon was a powerful force behind sweeping changes in management and policy, but it also is clear that deep complexity underlies the icon, and if solutions are not based on a rich understanding of forests and

1

how people relate to them, unintended consequences can generate ongoing social ferment.

In the second chapter, Johnson and Swanson set the stage for a broad historical sweep of worldviews and national forest policies, each of which places a marker for how society has viewed forests through time. They demonstrate that the idea of old growth has not been constant, and it is really only since 2003 that conservation of old growth has become a stated goal of a national policy (Healthy Forest Restoration Act). Despite its prior absence from national policies, old growth has been central to the ecology, economy, and social condition of the Pacific Northwest for centuries, dating back to the native peoples of the region. Their chapter presents clear evidence of the long-term social importance of old growth and the social and scientific diversity of the concept. Thus, they also propose that society's treatment of old-growth forests will continue to change in the future.

Chapter 1

Searching for Old Growth

THOMAS A. SPIES AND SALLY L. DUNCAN

The flat, glacial outwash plain in northern lower Michigan is savannah-like and open to the sky, but it was not always so. Only scatterings of large, gray, rotted stumps remind me that old-growth eastern white pine, red pine, and hemlock forests flourished here for hundreds of years. In 1972, almost 100 years after intensive logging of the area ceased, I (Spies) visited the area and found no large pines or hemlocks, just shrubs and patches of big tooth aspen and scrub oak. The remarkably slow recovery of the vegetation is thought to be a result of extensive logging followed by repeated, hot logging–slash fires that burned off the organic matter from already nutri-ent-poor, sandy soils. These activities effectively eliminated seed sources of pines and hemlocks that had developed through millennia.

A few hours later, I stopped at Hartwick Pines State Park, seeking to get some idea of the appearance of that long-gone old pine forest. Here, in an area smaller than fifty acres (it was eighty-five acres until a windstorm in 1940 knocked down almost half of the old growth), stands a remnant patch of white pines more than 300 years old, more than 150 feet tall, with a lower canopy of hemlocks. The park gave me some sense of what many of the northern Michigan forests might have looked like before Europeans arrived. But this small patch—just a fifty-acre speck in the former pine for-ests—with its well-worn trails and surrounding cutover landscape, was not much more than a botanical garden or curiosity.

Nevertheless, I was intrigued by the idea that millions of acres of Michigan and other states in the eastern United States, and much earlier in Europe, had supported old-growth forests of pine and hardwood whose dimensions we know only from written accounts of early Euro-American explorers and notes of surveyors. Most of the details of these forests—their appearance, structure, plant and animal composition, and pattern on the landscape—are left to our imagination, which may be filled with images from romantic American landscape painters, or even Walt Disney. In many ways, these forests are unknown—to all of us, scientifically and socially.

I did not know then that ten years later I'd be standing in the middle of the eroded volcanic mountains of the Pacific Northwest, surrounded by a region that still actually contained millions of acres of forests with massive old conifer trees—some 750 years old and eight feet in diameter, with crowns that disappeared into the mist more than 200 feet above me (fig. 1.1). Not only that, I would be the scientist, hired by Dr. Jerry Franklin of the U.S. Forest Service, the world authority on old-growth forests, to study the diversity in structure and composition of old-growth forests and to determine how different they were from younger natural and managed forests. Furthermore, whether I liked it or not, I was about to receive a fast-paced education in the increasingly wicked combination of the ecology, the management, and the politics of old-growth Douglas-fir forests in the region.

I began my old-growth research with a strong sense that many of the older forests of the Pacific Northwest might experience the same terminal fate as those of Michigan and other parts of the eastern United States. This feeling had a number of sources, not the least of which was sharing the narrow gravel roads with heavily loaded log trucks. It was reinforced every time I would take a five-year-old aerial photo out to a site to check an old-growth stand and see instead a recent clearcut and plantation of Douglas-fir seedlings. When I told people that I was studying old-growth forests, a typical reaction was to ask, "Why are you studying old-growth forests when they are just going to cut them all down anyway?"

Indeed, in the early 1980s, "they"—the Forest Service and Bureau of Land Management—had no old-growth conservation policy. The only protected old forests were those that happened to be in wilderness areas, which were typically in low productivity forest types, or in scattered, relatively small areas that were withdrawn from logging because of their steep slopes. Small protected areas, many of which were about the size of the state park in Michigan, contained some patches of old growth. I began to feel that I was documenting the region's remaining population of old-growth forests for

FIGURE 1.1. Old-growth Douglas-fir and western hemlock forest on the west slope of the Oregon Cascade Range. (Photo: E. Forsman)

posterity. In fact, odd as it now sounds, I saw myself and other scientists on the research project on a noble mission—a latter-day Corps of Discovery—to record and better understand the ecology of these poorly known forests so society could better understand its choices and so managers could use nature as a model in the design of managed forests and landscapes. Since the Lewis and Clark expedition to the region in the early 1800s, the amount of old growth on all lands in Oregon, Washington, and northern California had probably declined by sixty-five to ninety percent (depending on forest type, region, definition, and estimate of historical amounts) as a result of land clearing, timber harvest, and wildfire. This was our last chance to get it right: Our westward movement had, after all, run out of land, and all too soon we were seeing the end of what had at one point seemed like an endless natural bounty of old-growth trees.

I also quickly learned that my education about forests in the Pacific Northwest was only just beginning. For example, as debate on the future of old-growth forests began to spread, I discovered that I was by no means universally viewed as the guy in the white hat. This first struck me one night after I had finished giving a talk at the University of Oregon in Eugene on old-growth ecosystems. Following my talk, in which I described the parts and functions of old-growth Douglas-fir forests, one young man in the audience rose and echoed the poet Keats's complaint that Newton's experiments with prisms had destroyed the poetry of rainbows. He then asked, "Why do you scientists always have to dissect everything? Why can't you leave nature alone?" I was taken aback—wasn't the scientific approach the best way to address this problem? I responded with something about science being a way to help us understand and appreciate the mystery and beauty of these forests and a way to help us conserve them. I don't think I changed the mind of my critic, but his position made me ponder just what this old-growth debate was really about. It was my first clue that it was not always, or even very often, about the science.

My education about forest conservation in the Pacific Northwest really accelerated in the 1980s and 1990s. In March 1989, I was walking away from a symposium in downtown Portland where I had just presented some of the results of my old-growth research, when Jack Ward Thomas, later to become chief of the U.S. Forest Service, stopped me and asked me to join a team that was developing a management plan for the conservation of the northern spotted owl. I told him I'd think about it but later declined the opportunity because I had too much work to do analyzing all those old-growth data, and I did not have time. Policy issues, I thought, just took time away from real science. I now know that I was just about to be swept into the policy turmoil, setting me on a path that included more than scientific research. Just two months later, I'd be wearing a tie and explaining forest succession to Congress in Washington, D.C., at joint House subcommittee hearings on the management of old-growth forests of the Pacific Northwest.

Four years and several court injunctions later, much of the timber harvesting in the Pacific Northwest had been shut down by a U.S. judge. I found myself on the thirteenth floor of a bank building in Portland, with sixty-five other scientists who were part of one of the largest science and forest policy assessments ever attempted in the United States. We had been asked by President Clinton to develop, in just ninety days, a suite of scientifically based options for protecting old-growth forests and species of the region on federal lands. Under the Northwest Forest Plan that eventually emerged amid further controversy, conservation and development of old growth moved to center stage. Millions of acres of federal land in the Pacific Northwest are now dedicated to the protection and development of old-growth forests, including fire-prone types that present special conservation challenges (fig. 1.2). Logging of old growth on federal lands has slowed to a trickle—far below what the plan actually allows.

Now more than fifteen years have passed since the Northwest Forest Plan was established, arguably since the period when old growth as a

BOX 1.1. THE FOREST WAR

Frederick J. Swanson

The well-used metaphor of *war* aptly characterizes the conflict enveloping change in federal forestry in the Pacific Northwest in the late 1980s and early 1990s—the period of the forest war, the old-growth war. The public media, the courtroom, forest access roads, and even treetops where activists perched were the battlefields. Representatives of the forest industry; foresters; environmentalists; and elected officials of county, state, and federal governments were the principal combatants. They used the weapons of lawsuits, public demonstrations on city streets and forest roads, and loud voices wherever they might have desired effects. Old-growth forests, the spotted owl, photographs of forlorn children of loggers, and bumper stickers reading "Earth First! We'll log the other planets later" were among the banners in the battles. As with many wars, a set of latent tensions intensified until a triggering event erupted into open conflict, which raged and then subsided as combatants won and lost battles and general fatigue set in. Old growth was central to the run up to war and was a stage for the war itself. What will be the place of old growth in society in the developing postwar period? Can those who fought in this war move beyond seeing old-growth and forest issues as a battle? Could it be that new common "enemies"—fire, climate change, or globalization—change some former foes into allies?

regional and global icon truly "came to power." This change occurred not only on federal lands in the Pacific Northwest; the states of Oregon and Washington also have revised their plans to include protection of older forests, and some timber companies have issued statements swearing off the cutting of old-growth forests as part of forest certification. Outside the United States, in places such as British Columbia, southeastern Australia, and northern Sweden, where thousands of acres of old, pre-Industrial-age forest still exist, old growth also has been rediscovered by societies with new priorities for management. In the eastern United States and other places where little or no old growth exists, some are seeking to locate and protect the tiny fragments of old growth, let forests return to the wild, or "restore" old growth. However, the old-growth controversies in the Pacific Northwest have captured the public and scientific interest around the world. The global prominence of the Pacific Northwest on this issue may not be so much because of its uniqueness but because humans can easily see the broader implications of the struggle to protect the idea and physical reality of wild forests in an increasingly human-dominated world.

FIGURE 1.2. Old-growth ponderosa pine with a history of surface fire. (Photo: T. Iraci)

The centuries-old trees of the Pacific Northwest now exist in a new social, scientific, and economic world—different, certainly, from the heyday of westward expansion 100–150 years ago, different even from twenty years ago. What roads have led us as a society from the ghost forests of the eastern United States to the black hat/white hat schizophrenia of old-growth science and old-growth values in the Pacific Northwest? What have been the roles of science, policy, and politics in the spectrum of changes? It is wise perhaps to question whether the options developed by a group of scientists in ninety days fifteen years ago for the Northwest Forest Plan could really have provided a sound scientific and social basis to meet society's needs for forest ecosystem goods and services far into the future. What is the prognosis for our oldest forest stands and for the many other components of forest biological diversity, as we move toward an uncertain future? Will the pendulum swing back to the timber policies that led to removal of old growth in the eastern and central United States? Or does the concept of a pendulum swinging through a finite set of beliefs and options severely limit our understanding of what old growth really means?

What This Book Is About

Our goals for the book are to provide a broad suite of perspectives in chapter form on the changes that have occurred in the valuing and management of old-growth forests in the Pacific Northwest. We seek to provide the building blocks for a more comprehensive and in-depth understanding of the old-growth issue, drawing from diverse ecological, social, and economic perspectives, and trying always to recognize their complex interactions. Although many of the chapters are by academics, the intended audience is anyone who is interested in obtaining a better understanding of how people think about forest conservation and management with regard to old-growth forests. Consequently, the chapters do not assume a specialist's level of knowledge in these topics, and arguments within them are not overly burdened with jargon, data, and citations. These are informed opinions from a variety of people who have collectively thought about this issue for more than 500 person-years. We address several questions: Will the current path lead us where we want to go? What problems, paradoxes, road blocks, and uncertainties remain? What does the "new world" of climate change, globalization, and urbanization portend for the old trees and forests that many people in the Pacific Northwest now seek to conserve?

Our premise is that old-growth forests have become icons—objects of uncritical devotion that carry relatively fixed notions of what in fact is a dynamic and complicated mix of ecological understanding and human values. The enduring old-growth icon arose from a time of social turmoil beginning in the last quarter of the twentieth century and rapidly became entrenched as a most effective rallying cry against many aspects of forestry and public land management.

The old-growth icon takes many forms. All of them relate to how we view nature in general and forests in particular and sharpen as we understand how human impacts on the environment have accumulated momentum through time. The characteristics of this icon include

- *Naturalness*—a world apart from human intervention; a place were nature reaches full expression or even perfection
- *Stability*—the end of succession (climax); ancient trees whose age predates the arrival of Euro-Americans
- *Habitat for charismatic species*—the exclusive home of the northern spotted owl and other threatened and endangered organisms

- *A charismatic ecosystem*—a place of interconnectedness, high diversity, and ecosystem functions that relate to water use, carbon storage, and the reservoir of species with untapped potential value to society
- *Beauty*—aesthetic appeal—associated with appreciation of form, light, and colors that may derive from our European, artistic traditions
- *Place of wonder and mystery*—of spiritual experience, impervious to the world of science, and scientific mysteries; unknown species and relationships
- *Economic and aesthetic value of the product*—high value of large pieces of tight-grained wood that comes from old trees.

These iconic expressions have actively altered the way society relates to old-growth forests and have catalyzed major changes in policy and management. How appropriate they are and how well they really serve in reaching conservation goals for old growth are the questions we tackle in this book. How does our changing scientific knowledge of forests and their history square with the various iconic values of old growth? Do forest policies that were based on one set of scientific or social assumptions meet the needs of a changing world?

We wonder, for example, whether the power of the icon has unintentionally sown the seeds for the next forest crisis or loss of many of the values that were originally sought. Public perceptions and many conservation approaches have emphasized a static view and typically treated the problem as black and white: A forest is either old growth or not, the landscape should be divided into either old-growth reserves with no management or timber production areas with high-yield forestry. We may seek one or at most a few definitions of old growth, but more recent work in ecology has demonstrated how diverse forests and old growth are and how disturbances such as fire, wind, and insects are important to old-growth function and development. If we try to pin the definition down in place and time, might it ultimately elude us? Most challenging of all are fundamental differences in personal worldviews about old growth. These ecological and social complexities make it difficult to get consensus, conduct planning, and carry out management activities that are needed to help protect and restore native forest diversity and provide other forest values.

We examine the story of where we are with respect to old growth today in six parts. In part I, we lay out the story of old-growth values and conservation in the Pacific Northwest and demonstrate how it is about the ferment and evolution of ideas in ecology, management, economics, and society. Concepts of old growth emerge differently in different arenas through time, influencing each other directly, indirectly, and unevenly.

In part II, we explore old growth through the ideas of leading ecological and social scientists. This section sets up the idea that scientific perspectives on and knowledge about old growth continue to evolve in multifaceted ways, even though some aspects of the icon have stabilized around ideas from more than a quarter-century ago. These chapters raise serious questions about the fundamental basis of our ideas about old growth and our relationship with it.

In part III, we present different perspectives and different ways of expressing how humans have related to old growth in the immediate past and how those relationships affect our current and future old-growth policies. The authors were selected for their range of views and forms of expression and their depth of experience with forest issues in the Pacific Northwest.

In part IV, we address the implications for the future management of old-growth forests from the perspective of changes that have occurred in policies and global forestry. We also consider the barriers and opportunities to learning through adaptive management and new ways of thinking that could help bring economics into greater congruence with social values. Finally, we examine the different forms of uncertainty and ways in which decision-making processes can operate under uncertainty.

In part V, we consider how evolving science and social knowledge might be used on the ground, with actual trees and in real landscapes. We consider, for example, questions such as how do we manage at both stand and landscape levels and what new economically viable management strategies are available to private landowners seeking to contribute to the biodiversity that is often associated with the iconic old-growth forests.

Our concluding section presents two types of synthesis. The first focuses on themes that emerge from the different chapters—is there convergence among disciplines and worldviews? What are the roots of differences, what are the real implications of using icons, and what are some more effective ways to relate to forests than we do now, given the diversity of opinions on forest management? The second chapter in this section draws from the multiple perspectives presented by the authors to lay out some new ways of thinking about conservation of older forests, suggesting options for policy and management that might more effectively complement—or even override—our iconic views of old growth.

Chapter 2

Historical Context of Old-Growth Forests in the Pacific Northwest—Policy, Practices, and Competing Worldviews

K. NORMAN JOHNSON AND FREDERICK J. SWANSON

The history of our changing relations with the forest is painted on the landscape in the distribution of forests of various types and ages—old-growth forest is a critical part of that mosaic. It is important to put consideration of the future of old-growth forests in the context of their past. To set the stage for chapters that follow, we trace the history of management of Pacific Northwest forests: how we have perceived the forests at different points in time, the goals and constraints for use of the forests, and how we have actually treated the forests. These changing relations are expressed in our objectives, language, legislation, regulation, and policies and in the history of logging, fire suppression, and other management actions. The interplay of these factors in part represents the balance of power among competing worldviews of the value of forests—in the starkest contrast, for commodity exploitation and use versus their value as natural ecological systems.

Our focus will be on the conifer forests of Oregon and Washington, which we divide into wetter types (e.g., Douglas-fir, western hemlock) west of the crest of the Cascades and north of the Rogue River and drier types (e.g., ponderosa pine) that lie to the east and south of these geographic boundaries. We define *old growth* as forests of advanced age (at least 150–200 years old) that are characterized by structural complexity and a mix of species. We consider the roles of private, state, and federal lands in the conservation and use of old-growth forests. These ownerships have distinctive management objectives, rules, regulations, and practices. Most of our story, though, will inevitably focus on our federal forests, because they have the strongest commitment to grapple with the question of how to deal with old growth and that is where we now have the most remaining.

A Chronology

Native people lived with the forests of the Pacific Northwest for millennia — for the entire existence of modern/postglacial forests before Europeans arrived. But we have but little idea of how native people viewed old-growth forests. We do know that some of these forests, such as old-growth cedar groves, permeated the lives of coastal natives, who built homes of cedar planks, totem poles of whole tree boles, and baskets of bark. Also, understory plants were probably important for sustenance and medicines. These values continue to this day.

The effect of native people on forests appears to have ranged from profound to slight — profound in lowland valleys and foothills (well documented by David Douglas in journals from travels in 1820s; Whitlock and Knox, 2002) where fire was a management tool, and very minor in remote valleys in the Cascades well removed from major travel routes and mountain resources, such as berry fields and obsidian gathering sites. In the broad picture we feel that the influence of native people on the character and distribution of old-growth conifer forests of the region was localized in sites where other resources attracted use and that the frequency of fires observed in the paleoecological and dendrochronological records could have been generated by lightning strikes alone.

1785–1900 — Farms to the First Forest Reserves

As European settlers and government agents drove tribes of native people from their lands, land policies of the U.S. government gradually took shape and effect. Originally, land policies in the United States were designed to shift all land into private hands, and hundreds of millions of acres passed into private ownership from 1780 to 1880 (Dana and Fairfax, 1980). Few restrictions were placed on private land use. In many places forests were converted to farms: From 1780 to 1880, more than thirty percent of the forests in the East and South were converted to cropland, and the white pine in the upper Midwest was liquidated by the late 1800s.

Early Euro-American settlers (mid-nineteenth century) in the Pacific Northwest likely saw old-growth forests as an impediment to travel, agricultural development, grazing, and perhaps other uses of the landscape, but like the native people, there was probably little overlap of primary land occupancy and old-growth forests. The first significant encounter between western civilization and old growth may have come with initial logging powered by oxen, steam, trains, and human sweat—this was the same kind of big timber featured in Paul Bunyan stories of the logging of the upper Midwest. This early exploitation of the timber resource occurred mostly on private lowlands forests that had been carved out of the public domain. This harvest was first consumed locally for home building and other uses, but the California gold rush created a commercial market for the wood from Northwest forests, as did the building of railroad lines across the West.

In the mid- and late 1800s, forest policy became a national issue as cutover forests became common and Congress established the first forest reserves and passed a series of acts focused on the role of forests in the West (Dana and Fairfax, 1980). This period culminated with the Organic Act, which established the purpose of the reserves and authorized the sale of large, dead, and mature trees, viewing old-growth trees primarily as wood products (table 2.1).

1900–1960—Rise of Sustained Yield Forestry

As with most laws, the Organic Act left much room for interpretation. Some hoped that the protection aspects of the language would be emphasized; others hoped that the use aspects would be highlighted. Gifford Pinchot, the first professionally trained forester in the United States and soon to be the first chief of the U.S. Forest Service, came down strongly on the side of use, albeit conservative use, focused on contributing to the economic development of the West under the supervision of resource professionals. Pinchot's vision led to the sustained yield model that dominated forestry for most of the twentieth century. The forest, including water and forage, was to be used at a rate that would not impair its permanent value (Dana and Fairfax, 1980; Johnson, 2007).

This vision was to be implemented through scientifically trained foresters with their "science" coming from disciplines that supported the management paradigm of the day—the sustained yield model of forest management. Forests would be grown as a crop with the goal of a "regulated forest" in which the same amount of timber would mature and be harvested each year. Maximum harvest age would generally not be greater than the age of maximum average growth—eighty years for Douglas-fir on a medium productivity site, which is well short of old growth—and it could be shortened for economic reasons. Wildfire and pests would be controlled and suppressed. All lands that could produce commercial crops would produce them over time. Forest lands would be roaded to provide access for timber harvest and to allow control of wildfire. Old growth—"decadent, over-mature, slow-growth forests"—would be a high priority for harvest to make way for fast-growing, second-growth stands and as a source of revenue and high-quality wood products (Cleary, 1986; Hirt, 1994; Langston, 2005; Johnson, 2007).

By World War II, much of the high-quality old growth on private lands in the Northwest and across the West had been cut. Attention then turned to public lands, especially the national forests; the sustained yield model became the reigning management system for these lands. The Forest Service was ready, having assessed the federal forest resources of the region, with old growth described and mapped as "large saw timber," reflecting the commodity value of this forest.

Clearcutting dominated in the federal forests in wetter forests west of the crest of the Cascade Range (Westside) in Oregon and Washington. *Clearcutting* was termed "even-aged management" and represented to the public in corporate advertisements as nature's way—like stand replacement fire, but without wasting the wood. In the drier pine forests to the east of the crest of the Cascades (Eastside), single-tree selection was used to target the big, old ponderosa pines that were most susceptible to insect attack. In both cases, harvest targeted old-growth stands and trees for removal, and this action was justified in the name of forest "health" or removal of "decadent" old trees.

As allowable cuts on the national forests shifted from theory to reality in the 1950s to meet the demands of the postwar housing boom, the Forest Service was forced to reduce existing primitive-area designations, which in turn intensified the call for formal wilderness, leading to the Wilderness Act of 1964. However, few commercial timberlands were included in the designated areas. The act also called for the president to identify other roadless areas (one-third of the national forests) to Congress within a decade. This provision became the next political battleground over federal forest management and had the net effect of delaying entry into many of these areas for decades.

TABLE 2.1. Major federal legislation relevant to forestry and especially old-growth forests.

ACT	DATE	PURPOSE	OLD-GROWTH FORESTS RECOGNIZED AS A VALUE?	EFFECT ON OLD-GROWTH USE AND CONSERVATION
Land disposition acts	1785–1860	Shift public domain into private hands	No	Shifted most of the low-elevation, productive old growth into private hands, where it was harvested
Creative Act	1891	Allow president to retain parts of public domain	No	Enabled incidental retention of some old-growth forests
Organic Act	1897	State purposes of forest reserves established by president: continuous supply of timber and protection of watersheds; allow sale of large, dead, or mature trees	No	Allowed old-growth forests to be harvested for commercial purposes; limited harvest to large, dead, or mature trees
O and C Act	1937	Establish goals of BLM lands in western Oregon; provide sustained yield of timber and contribute to community stability	No	Committed old-growth forests on O & C lands to harvest as part of sustained yield
Multiple-Use Sustained-Yield Act	1960	Recognize goals for which national forests were being managed	No	None

Act	Year	Purpose	Legislative protection	Effect on old-growth forests
Wilderness Act	1964	Maintain areas in a wild state forever; establish process for considering other federal roadless areas for wilderness	No	Some old-growth forests incidentally placed within wilderness areas in a permanent reserve status
Endangered Species Act	1973	Prevent further harm to species listed as threatened or endangered and their habitat; make recovery of species a primary objective for federal lands	No	Provided a legal mechanism for conservation of old-growth forests—as habitat for endangered species, especially on federal lands
National Forest Management Act	1976	Require integrated planning; put a variety of controls on timber harvest; provide for diversity of plant and animal communities	No	Provided a legal mechanism for conservation of old-growth forests on the national forests—as a contributor to the viability of designated species; removed Organic Act requirement that harvest be limited to large, dead, or mature trees
Healthy Forest Restoration Act	2003	Expedite actions to reduce fuel hazard on federal forests while maintaining or contributing to restoration of the structure and composition of old-growth stands	Yes	Provided the first legislative protection for old-growth trees and forests

Passage of the Multiple-Use Sustained-Yield Act in 1960 did little to change the worldview inherent in the sustained yield model and conversion of old forests to tree plantations. Timber primacy still drove federal agency priorities (Cleary, 1986), with foresters in charge, tremendous demand for the softwood timber on the national forests, and the continuing concern over timber scarcity. But now federal forest managers targeted harvest of high-value, old-growth forest to capitalize on the clear, tight-grained wood and to get "thrifty young stands" growing in place of the "decadent, over-mature forest."

1970–1990—Emergence of Ecological Forestry

In the early 1970s, three laws were passed that changed the power balance in the debate over national forest management, where most remaining old-growth forest existed in the Northwest. Those laws were the National Environmental Policy Act (NEPA) in 1970, Endangered Species Act (ESA) in 1973, and National Forest Management Act (NFMA) in 1976 (table 2.1). Although neither NEPA nor ESA specifically mentioned old-growth forests, their focus on assessing environmental impacts and on conserving ecosystems would inevitably have implications for the management of the remaining old-growth forests.

The NFMA originated in the clear-cutting controversy on the national forests, which began in West Virginia, Montana, and Wyoming in the 1960s and became a national issue in the 1970s. The NFMA reaffirmed that the national forests should be managed for multiple use and sustained yield. The NFMA, however, among other things, called for national forest management to provide for diversity of plant and animal communities based on the suitability and capability of the specific land area in order to meet overall multiple-use objectives. In turn, the regulations implementing NFMA called for fish and wildlife habitat to be managed to maintain viable populations of vertebrate species in the planning area. A viable population was regarded as one having sufficient reproductive individuals to ensure its continued, well-distributed existence in the planning area. Taken together, NFMA and ESA embodied a strong, affirmative responsibility on the part of a land management agency to ensure the viability of species dependent upon habitats found in the national forests and grasslands or, if they were threatened or endangered, to ensure that they be recovered to the standard of viable populations.

In this way, forest planning and policy for the national forests began to slip away from the framework of sustained yield and economics and toward that of ecology and conservation biology. It was a critical turning point for the future of old-growth forests.

While these social perspectives and policies were evolving in the 1970s, so was the science of ecology and forestry. Whole new branches of ecology were developing, including landscape ecology and conservation biology. The Forest Service adopted ecosystem management—a more holistic ecological and social approach to forest management goals and practices—in the 1990s. A central axiom of the new ecological approach to forestry was stated by Seymour and Hunter (1999: 29): "Manipulation of a forest ecosystem should work within the limits established by natural disturbance patterns prior to extensive human alteration of the landscape." The key assumption here is that native species were adapted to these circumstances and perhaps in some cases even dependent on the natural disturbance regime. Thus, maintaining a full range of similar conditions under management would offer the best assurance against losses of biodiversity (Johnson, 2007).

Ecological studies in the Pacific Northwest Blue Mountains and H. J. Andrews Experimental Forest and the Forest Service research effort "old-growth wildlife habitat research program" helped set the stage for the rise of old growth and the role of scientists in the policy arena. Increasingly, old growth became a central focus in the eyes of the public as the fight over exploitation versus protection erupted, pitting people who saw old growth as "over-mature timber" against those who saw "cathedral-like, ancient forests" (see chapter 11). Despite the lack of direct legislative recognition of old-growth forests (table 2.1), these forests increasingly became the political focus of ecological management.

1990–2000—The Final Battle

CONSERVATION OF BIODIVERSITY AS THE ASCENDANT GOAL
ON THE FEDERAL FORESTS

By the late 1980s, the science, laws, policies, and economic and social forces were commencing a major battle about the management of the Northwest's federal forests in general and the management of old forests in particular. The triggering event occurred when environmental groups sued the Forest

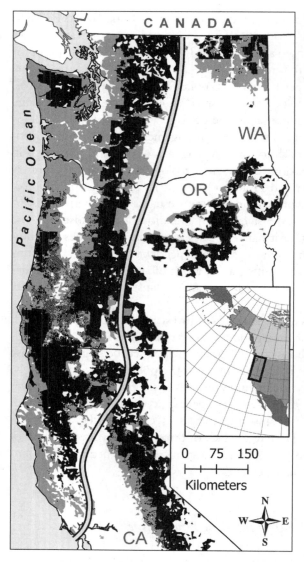

FIGURE 2.1. Federal forest lands (black) and nonfederal forest lands (gray) in Washington, Oregon, and northern California. Federal lands to the west of the double line fall within the range of the northern spotted owl and the Northwest Forest Plan.

Service over the adequacy of protection for a wide-ranging species that favored *late-successional* (older forests, including old growth) forests—the northern spotted owl. In the pivotal lawsuit, Seattle Audubon Society in 1989 sued the Forest Service for failing to adopt a credible conservation strategy for the northern spotted owl that would comply simultaneously with NEPA, ESA, and NFMA (Caldwell et al., 1994). The courts agreed with the plaintiffs and issued a restraining order on the harvest of "owl habitat" until a

BOX 2.1. THE NORTHWEST FOREST PLAN FOR FEDERAL LANDS

Thomas A. Spies

The Northwest Forest Plan of 1994 ushered in a new era of forest management on twenty-four million acres of federal lands (Forest Service and Bureau of Land Management) across three states of the Pacific Northwest (fig. 2.1). Bounded by the range of the northern spotted owl, the plan sought to protect and restore old-growth forests and watershed and streamside conditions to maintain viable populations of birds, mammals, fish, amphibians, plants, fungi, and lichens whose habitat had been reduced by more than 150 years of logging, development, wildfire, and its suppression. In addition to protecting native biodiversity, the plan called for producing a predictable and sustainable level of timber sales from federal forests.

The plan attempted to meet these goals by dividing the twenty-four million acres of federal forest land (nearly half of all public and private lands in the region) into zones or allocations including "late-successional reserves" (thirty percent of plan area), "riparian reserves" (eleven percent of plan area), "adaptive management areas" (six percent of plan area), and "matrix" lands (sixteen percent of plan area) where timber production could be carried out. The remaining thirty-seven percent of the land was already zoned into areas excluded from timber production, such as wilderness and administratively withdrawn land. Under these allocations, eighty percent of the existing mature (roughly eighty to 200 years old) and old-growth forests (large trees, including some more than 200 years in age), or 2.4–6.4 million acres, depending on the definition, were protected in new or existing reserves. Twenty percent of mature and old growth occurred in the matrix and was expected to be a major source of revenue from timber harvesting. Logging in the matrix was to be done using new ecologically based silviculture practices that left patches of trees to retain species and habitat diversity instead of removing all of the trees in a clearcut.

The plan also called for thinning in the numerous plantations that occurred inside reserves. The purpose of this was to restore ecological diversity and facilitate future development of old growth forests and spotted owl habitat. In drier forests thinning and prescribed burning were allowed in older forests in late-successional reserves to reduce fuel in the form of live and dead shade-tolerant conifers and shrubs that had accumulated as a result of fire suppression. In the historical natural fire regime, these forests had been regularly thinned out by frequent low- to moderate-severity fires.

Monitoring and adaptive management were expected to be important parts of the plan. A decade into implementation of the plan, monitoring revealed these outcomes:

- There was a net increase in the extent of older forest—most of it a result of incremental growth of younger forests into larger tree diameters and less logging of old growth than originally planned.

- Losses of old forest to high-severity wildfire were less than expected across the region, but in drier subregions losses were so high that much of the older forest will be gone in fifty years, if recent trends continue or accelerate.

- The area of dry forests thinned to reduce fire hazard was far less than thought to be needed to reduce risk of loss to high-severity fire.

- Many spotted owl populations continued to decline, despite the steep reduction in cutting of old forests. It is widely believed that range expansion by the more aggressive barred owl may be a cause.

- Timber production was much lower than expected because of continued social pressure against logging of old trees anywhere, even those outside the scientifically designated reserves.

The plan's adaptive management process, designed to learn about new ways to integrate forest management objectives and deal with uncertainties, was not widely or aggressively implemented right from the start. Uncertainty, both ecological and social, and lack of trust remain some of the greatest challenges to the plan and to coherent old-growth management across the region.

credible strategy could be developed to conserve the northern spotted owl (see chapter 17).

The political, policy, and science spheres lined up, leading to unprecedented direct involvement of scientists in the policy arena (Franklin, 1995). These efforts included the Interagency Scientific Committee report on the northern spotted owl (Thomas et al. 1990), the "Gang of Four" (Johnson et al., 1991), and "FEMAT" (Forest Ecosystem Management Assessment Team) (FEMAT, 1993). They culminated in the Northwest Forest Plan (NWFP) (box 2.1), which continues to guide forest policies on the federal forests of the Pacific Northwest. Collectively these strategies turned protection of species associated with old-growth forests into a primary goal for forest management—timber management would come second. They also recognized that old-growth conservation was about more than owls and always had been: The NWFP calls for conservation of more than 1,000 species thought to inhabit old-growth forests.

The NWFP strategies were estimated to allow harvest at approximately twenty percent of historical levels—with much of that reduced harvest still

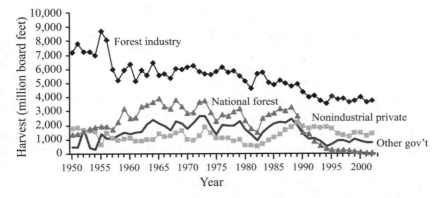

FIGURE 2.2. Harvest for the Douglas-fir region (western Oregon and Washington).

coming from old-growth forests. As the NWFP was implemented, even that modest level of old-growth harvest generally did not occur for several reasons, including public protest, litigation, and attempts to protect little known species (Thomas et al., 2006).

In the Eastside ponderosa pine region outside the NWFP area, similar changes were afoot (fig. 2.1). Under the threat of a lawsuit, the Forest Service instituted a temporary measure to conserve the remaining large trees (greater than twenty-one inches in diameter) and old-growth stands, largely reversing the direction of the national forest plans they had just put in place and halting the harvest of large, old trees on the Eastside national forests. That temporary measure is still in effect.

These new policies and their implementation led to the collapse of the old-growth timber economy (see chapter 7). By the 1990s, harvest on private lands came almost exclusively from young growth that had regenerated after the virgin stands had been cut (fig. 2.2). That harvest continued largely unaffected by the debate over endangered species and old-growth forests. On the federal lands, however, old growth was the dominant source of wood and the halting of old-growth harvesting accelerated the closure of old-growth mills and the retooling of others to handle small logs from private lands. The decline in harvest also forced wood products workers to shift to other, often lower paying, jobs and caused significant difficulties for some communities (Helvoigt et al., 2003; Thomas et al., 2006). Even with this contraction of federal harvest and hardship to people and individual communities, the economy of the Pacific Northwest, as a whole, has continued to grow. Thus, federal harvest levels have shifted from a regional economic problem to a local, community problem, and national political energies have largely turned elsewhere.

An additional issue surrounds the "O and C" lands managed by the Bureau of Land Management in western Oregon (table 2.1). Originally given to the Oregon and California Railroad Company in payment for construction of a railroad connecting the two states, the federal government took back these lands when the railroad violated terms of the deed. In the mid-1930s, they were given a special mandate to provide a permanent timber supply and contribute to economic stability of local communities. The NWFP treated these lands the same as the national forests. A lawsuit challenged that interpretation and the Bureau of Land Management to develop a new management plan with their own unique approach, perhaps somewhat similar to the state plans discussed below. The future of old-growth forests under that plan is unknown: Levels of cutting of old growth may increase on those lands, although not back to harvest levels of the late twentieth century.

EFFORTS ON STATE LANDS

State forests in both Oregon and Washington cover large areas, especially in the coastal provinces. In the past ten years, state forest management agencies have attempted to merge the sustained yield model with the ecological forestry model: providing a steady income from timber products while recognizing the ecological values of old forests and the species within them. Washington State has committed to conserving its remaining old forest while agreeing to a long-term Habitat Conservation Plan for endangered species and has also drawn up a new site-specific management plan to guide harvests. Oregon, which has mainly young forests established after major fires in the nineteenth century and early to mid-twentieth century, has committed to a "structure-based" management approach that will provide more diverse forests in the future while attempting to provide a steady stream of revenue from those lands.

2000—Old-Growth Forests, Fire, and Fuel

The sustained yield model of forestry generally called for the suppression of all wildfires that occurred. In dry forests, the resulting buildup of stand densities, compounded by the removal of the large fire-resistant ponderosa pines, resulted in a major fuel management problem. This change potentially converts an ecosystem characterized by high-frequency, low-severity fires to forest conditions susceptible to severe, high-intensity, stand-replacing fires.

The fires across the West in the early years of the current century, culminating in the San Bernardino fire of 2003, created intense presidential and congressional interest. Thus, the call for legislation to expedite fuel treatments across the West led to the passage of the Healthy Forest Restoration Act—the most important piece of forest conservation legislation in the last quarter-century. That law expresses congressional intent and desire to speed up fuel treatments, requires that danger of inaction be considered equally with danger of action in NEPA analyses, and limits the time judges have to consider attempts to stay actions. The Healthy Forest Restoration Act set two important precedents: It provided (i) the first statutory protection of old-growth forests and (ii) the first statutory call for the use of past ecological conditions in planning.

Old-Growth Forests of the Pacific Northwest in a Global Context

Some might view our changing relations with old forests in the Pacific Northwest as globally unique but, in fact, the region's struggles can be seen as just another manifestation of a global process in which human cultures have shifted their engagements with forests often and in similar ways. Many preindustrial cultures inhabited old forests, and other cultures avoided the "dark forests" where that was possible. As the number of people and their capacity to change the environment grew, they cleared forests for agriculture and grazing. At times, this land use conversion stage of human engagement gave way to unregulated exploitation. With elimination of forests came growing concerns about future timber famines and the environmental impacts of harvests. These fears led to regulations to constrain and guide forest management in many parts of the developed world.

The timing and terminology of these forests conservation efforts differed, however, by culture and ecosystem and, as in the United States, the policies often did not explicitly recognize "old growth." For example, in Australia the focus of forest conservation efforts that began twenty years ago was on "rain forests"—a complex forest ecosystem defined not so much by successional stage as by species composition, structure, and climate. In Germany and Austria, where humans have used almost every acre of forest over the past several thousand years, some small fragments of "virgin" or "primeval" forests (lacking evidence of prior human uses for logging or grazing) are left in the mountains, and some of these were protected by religious communities. In the past twenty years in Europe, many areas of previously used forests have been put into reserves that will become "virgin forests of the future," according to some European ecologists. It may be that intentional culturing or "restoring" of old-growth forests is now part of the story—perhaps for the first time in history.

Thus, the social and policy trends we see in the Pacific Northwest appear to be part of a more general pattern of treatment of iconic, complex natural forests—however named: habitation or avoidance → exploitation → conservation → cultivation—but the sequence has varied from region to region depending on the timing of initial encounter by industrial societies. Currently, in parts of the developing world, old, native forests (especially rain forests) are undergoing unregulated exploitation. The Pacific Northwest of North America has experienced full expression of this sequence in part because the forests were extensive and European invasion was recent and incomplete, as also is the case in Australia and Canada (Lane and McDonald, 2002). In this region, we can see the pattern of shifting human engagement with the forest in the history of language for old forests: The "virgin, old-growth" forest came to be viewed as "large saw timber" in the exploitation stage (first half of the twentieth century) and then they became "ancient forests" in the 1980s–1990s conservation era. How will we speak of the old forests in coming decades? Will the pattern of our treatment of these forests become more nuanced, less linear, and more varied in space and time?

Conclusion

Old-growth forests have been central to forest management opportunities and debates throughout much of the history of American forestry. Our changing view of old growth is written in the social, policy, scientific, and forest management histories of our country. Despite this long history, it is only since the 1980s that the value of old growth has been recognized in forest plans and only since 2003 that its protection has been mandated in federal forest policy. The rise in the status of old growth and its naming do not mean, however, that the issue is now "solved" or laid to rest. If fact, we are only just beginning to understand what this sea change in forestry means. Old difficulties remain, and new ones are emerging. The concept of old growth still remains socially and scientifically diverse, making policy and management of it on the ground very challenging. In addition, new understanding of the importance of natural disturbances in the development of old growth and the effects of climate change challenge its iconic status as a stable, enduring entity. For example, debates over salvaging old forest after a fire have now sprung up. The public generally supports the idea of salvaging followed by planting, but from an ecological standpoint, such natural disturbances create distinctive habitat and provide a way of extending the ecological influences of old growth from one forest generation to the next. How old growth will be valued by the next generation of humans is unknown. However, if the past is any guide, a new chapter in the story of old growth will be written.

LITERATURE CITED

Caldwell, L., C. Wilkinson, and M. Shannon. 1994. Making ecosystem policy: Three decades of change. *Journal of Forestry* 92(4):7–10.

Cleary, D. 1986. *Timber and the Forest Service.* Lawrence: University Press of Kansas.

Dana, S., and S. Fairfax. 1980. *Forest and range policy,* 2nd ed. New York: McGraw-Hill.

FEMAT (Forest Ecosystem Management Assessment Team). 1993. *Forest ecosystem management: An ecological, economic, and social assessment.* Report of the Forest Ecosystem Management Assessment Team. 1993–793–071. Washington, D.C.: U.S. Government Printing Office.

Franklin, J. F. F. 1995. Scientists in wonderland. *BioScience Supplement* 45(6):S74–S78.

Helvoigt, T., D. M. Adams, and A. Ayre. 2003. Employment transitions in Oregon's wood products sector during the 1990's. *Journal of Forestry* 101(4): 42–6.

Hirt, P. W. 1994. *A conspiracy of optimism: Management of the national forests since World War Two.* Lincoln: University of Nebraska Press.

Johnson, K. N. 2007. Will linking science to policy lead to sustainable forestry? Lessons from the federal forests of the United States. In *Sustainable forestry—From monitoring and modelling to knowledge management and policy science,* edited by K. Reynolds, M. Shannon, M. Kohl, D. Ray, and K. Rennolls. Cambridge, MA: CABI.

Johnson, K. N., J. F. Franklin, J. W. Thomas, and J. Gordon. 1991. *Alternatives for management of late-successional forests of the Pacific Northwest.* A Report to the Agriculture Committee and the Merchant Marine Committee of the U.S. House of Representatives.

Lane, M. B., and G. McDonald. 2002. Towards a general model of forest management through time: Evidence from Australia, USA, and Canada. *Land Use Policy* 19:193–206.

Langston, N. 2005. Resource management as a democratic process: Adaptive management on federal lands. In *Communities and forests: Where people meet the land*, edited by R. Lee and D. Field. Corvallis: Oregon State University Press.

Seymour, R., and M. Hunter. 1999. Principles of ecological forestry. In *Managing biodiversity in forested ecosystems*, edited by M. Hunter. Cambridge, England: Cambridge University Press.

Thomas, J. W., E. D. Forsman, J. B. Lint, E. C. Meslow, B. R. Noon, and J. Verner. 1990. A conservation strategy for the northern spotted owl: a report of the Interagency Scientific Committee to address conservation of the northern spotted owl. USDA Forest Service, USDI Bureau of Land Management; Fish and Wildlife Service, Portland, OR. Washington, D.C.: U.S. Government Printing Office.

Thomas, J. W., J. F. Franklin, J. Gordon, and K. N. Johnson. 2006. The Northwest Forest Plan: Origins, components, implementation experience, and suggestions for change. *Conservation Biology* 20(2):277–87.

Whitlock, C., and M. A. Knox. 2002. Prehistoric burning in the Pacific Northwest: Human versus climate influences. In *Fire, native peoples, and the natural landscape*, edited by T. Vale. Washington, D.C.: Island Press.

PART II

Exploring Old Growth Through Ecological and Social Sciences

One distinctive feature of the old-growth controversy in the Pacific Northwest is the degree to which scientists played a central role in the policy debates, which for most of them were uncharted waters. This section provides a sample of their perspectives, some of which were once or still are controversial. The chapters progress from ecological to social sciences, reflecting a trend in the study of old growth. Seen in the early 1980s as primarily a subject for ecological science, the old-growth "problem" was framed increasingly as a social issue and thus a subject for social sciences research.

We propose in this section that scientific perspectives on and knowledge about old growth continue to evolve in multifaceted ways, even though some aspects of the icon have stabilized around ideas from more than a quarter-century ago. The most dramatic example has been our improved understanding of forest dynamics, which raises important questions about the long-term value and even the validity of geographically static reserves as our central old-growth preservation strategy.

In his chapter, Spies considers the science of old growth largely through the lens of vegetation structure and dynamics, illustrating the challenges in studying and understanding a complex ecosystem that changes through space and time. He lays the foundation for Noon and Forsman to examine old growth as habitat and to consider how other successional stages and processes affect wildlife. Reeves and Bisson tackle the challenge of putting fish into the old-growth story—How does an organism that spends all its time in water relate to old-growth forests? The answer is not so simple. Carey addresses the problem of understanding how forest management affects terrestrial biodiversity, exploring research about maintaining native biodiversity in managed forests. Haynes lays out the sweep of change in the economics of old growth over the past 50 years and illustrates how the economics of forests and personal values today differ from those current in the heyday of old-growth logging. Lee examines how modern society has come to value trees as sacred in an increasingly commercial world. Proctor continues with the theme of spirituality and examines the ambivalent and complicated relationships among environmentalists, science, and religion. Finally, Steel places the entire issue in the context of public discourse and politics, revealing how politicians use symbols to simplify complex issues and suggesting that solutions lie outside the sphere of advocacy democracy.

Chapter 3

Science of Old Growth, or a Journey into Wonderland

THOMAS A. SPIES

Old-growth forests often are valued for their apparent stability—they are often perceived to be ancient but ageless, unchanging through millennia. This notion of inherent stability is at least in part responsible for our desire to preserve such forests. Paradoxically, however, the scientific value of old growth really lies in helping us understand how forests change across in time and space. Old-growth forests are rich in history and constantly changing in subtle or not-so-subtle ways. Understanding the ecological complexity behind the old-growth icon may better prepare us to conserve forest biodiversity and old growth in the future.

In this chapter, I make the following points:

- Old growth is dynamic, resulting from the opposing forces of biological growth on one hand and disturbance and decay on the other.
- The scientific value of old growth lies in several areas: (i) providing controls for measuring the effects of human activities; (ii) shifting our focus to relatively long timeframes to help us understand how and why forests change; (iii) helping us identify the unique contributions of all forest stages to biological diversity and ecological processes; and (iv) opening our eyes to the importance of structural

complexity in providing habitat for organisms and the foundation for ecological processes.

- The conceptual and methodological challenges in the study of old growth include scale dependence, complexity, reification (making abstract concepts concrete), and the difficulty of conducting centuries-long controlled experiments.

- Our scientific understanding is evolving from simple models of forest development to more complex ones that do a better job of representing the dynamic maze that defines forest development. As our scientific understanding improves, it becomes increasingly clear that an old-growth icon—based on a well-defined, finely focused snapshot of stable forest conditions—is inadequate. It may even be a conceptual trap that keeps us from more broadly understanding the variety of ways in which forests develop, thus eroding our ability to provide for forest diversity in the long run.

My overall thesis is that in following our curiosity about tall, massive old-growth Douglas-fir forests we have been led down a rabbit hole into a world of ecological complexity filled with paradoxes (e.g., disturbance is needed to maintain old forest diversity) and visions defined by the different scales at which we view old growth. In this new world, old growth is only one part of the story of forest complexity, and in some views of the forest it may disappear except for the legacies of dead wood it leaves, like the smile on the disappearing face of the Cheshire cat in Lewis Carroll's *Alice in Wonderland*. Old growth remains vital to biological diversity, but our continuing investigation of it reveals that it is only one part of a complex and changing web of forest diversity.

Definitions

"Old growth" has almost as many definitions as "forest." In recent years forest ecologists have defined *old growth* as a forest in the later stages of development characterized by the presence of old trees and structural diversity. Of course, those later stages occur at different time points for different tree species, further complicating our attempts at definition. A second and somewhat older meaning is a forest that has developed without evidence of human impacts such as cutting or grazing. This second meaning, which also is associated with the words *virgin* or *primary*, is problematic from a scientific standpoint because it often is difficult to determine the human history of a forest and many seemingly virgin forests, have been influenced directly by aboriginal people or indirectly by industrial-age people through fire suppression or other activities. One ecologist aptly noted that "in forestry virginity is relative" (Clark, 1996).

Of course, the term *old growth* has its own history of meanings and values. Coined as early as 1891 by Bernard Fernow, an early leader in American forestry (Busby, 2002), old growth was simply a forest of old trees—not unlike the core meaning today. The addition of "growth," which is not really essential to the term's scientific meaning, may have occurred as a convention of a time when forests were primarily viewed as a crop to be harvested. When T. T. Munger, the first director of the U.S. Forest Service's Pacific Northwest Research Station, published a paper in 1930 on conversion of old forests to young, he did not include the term *growth*, but he did refer to the big trees of the region as "colossi of the vegetable kingdom," a choice of words that conjures up images of pumpkins, as old ponderosa pine are sometimes called by foresters (Munger, 1930).

Old-Growth Science in the Pacific Northwest

The science of old-growth forests in the Pacific Northwest was advanced in the late twentieth century by the publication of two important synthesis volumes. The first, which focused on the dry forests of the Blue Mountains of Oregon (Thomas, 1979), identified old growth as one of several successional stages that were important to wildlife species. Old growth was defined structurally in terms of live trees, standing and down dead trees, and presence of cover and canopy layers. Prior to publication of this book some wildlife biologists viewed old growth as a biological desert, because it was perceived to lack the habitat for game species.

The second synthesis (Franklin et al., 1981) characterized the composition, structure and function, and management options for old-growth Douglas-fir forests in the relatively wet western Cascades. This work called attention to the conservation status of old-growth forests—they were protected in only five percent of the landscape in parks and wilderness areas at that time. Franklin et al. (1981) also laid out a fundamental scientific question: How are old-growth forests (and their streams) distinguished from second-growth forests that follow fire or timber management?

This publication made sixteen major conclusions, including the following:

1) It takes 175–250 years for Douglas-fir forests to develop the range of structures associated with old growth;
2) Few plant and animal species are confined to old growth, but some may be dependent on it;
3) Net forest growth slows to near zero in old growth because mortality of trees is generally balanced by growth of surviving and new trees;
4) Old-growth forests hold on tightly to nutrients, and losses of nutrients such as nitrogen are low;
5) The structure of old forests is more heterogeneous than that of young forests;
6) The most distinctive features of old growth are large live trees, large snags, and large fallen trees on land and in streams;
7) The live and dead structures of old-growth forests provide specialized habitats for a variety of vertebrates, invertebrates, mosses, lichens, and fungi on land as well as in streams;
8) Conservation of old growth should be based on protecting entire old-growth watersheds from logging and, where timber management is practiced, leaving stream buffers, old trees, and dead wood within the managed forest.

These statements, which were not initially based on a large body of empirical science, are still largely valid after twenty-five years of subsequent research. However, many questions remain unanswered. For example, we still do not fully understand why the gross growth or production part of the forest ecosystem declines as forests age (Binkley, 2004), or why some species are more common in older forests than in younger ones.

Challenge of Scale

Two fundamental questions lie at the root of the ecological science of old-growth forests: (i) How do forests change in structure and species as they mature? (ii) What are the causes of those changes?

I will examine these questions further, but first, it is important to point out that the answers to them, like many in ecology and other sciences, depend on temporal and spatial scale. For example, the trees species that constitute the old-growth forests of the western Cascades appeared together for the first time only about 6,000 years ago—making old-growth Douglas-fir–western hemlock a *young* forest association in geological time. If we take a more limited view, say the 1,000 years prior to 1850, the species composition of these forests has been relatively constant but the amount of young and older forest has fluctuated over time as fires (some of them set intentionally by Native Americans) and other disturbances occasionally killed the canopy trees (fig. 3.1).

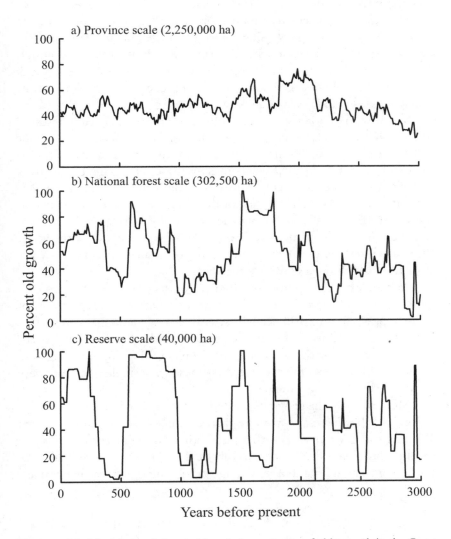

FIGURE 3.1. Modeled variation in historical percentage of old growth in the Coast Range at different spatial scales: (a) the area of the entire Coast Range physiographic province, (b) an area the size of a national forest, (c) an area the size of an old-growth reserve. (Wimberley et al., 2000)

During the past 100 years, forest change has been more unidirectional as the area of forests declined, replaced by open land or relatively uniform younger forests as a result of widespread land clearing, logging, and human-caused wildfires. During this same time in dry landscapes of the region, cutting of large old pine and suppression of fire has created denser, more fire-susceptible stands with fewer of the large fire-resistant pines.

Unfortunately, science cannot determine the "best" space or time scale at which to view forests. No single scale is sufficient to understand these natural systems. Forest patterns and dynamics are always varying (although not uniformly) across time and space. Our choice of a reference frame is driven by the scale of species or ecological processes, economics, institutional mandates, and personal interests.

Scientists have long struggled with scale. For example, early research in Oregon used small sample plots (less than one-tenth of an acre)—about half the size of a tennis court—to characterize the structure and composition of old growth. The plots were subjectively located around big old trees and avoided patches of recent disturbance from wind and fire. The scientists quickly realized that the scale of the old-growth forest phenomenon was larger than a small patch of big, old trees. With a small plot the death of single big tree could change the forest from old to young. Over the years, the size of forest sample plots has increased more than twentyfold to greater than 2.5 acres, the area of two football fields.

At the next scale up, foresters and ecologists often use the term old-growth *stand*. The concept of a forest stand, a relatively uniform area of forest typically ten to 100 acres in size, does not come from any scientific analysis or theory, but from the practical needs of foresters and scientists who needed relatively homogeneous units for logging, planting, and experimental study. At the stand level the death of a single tree is not very noticeable by most measures.

More recently, the geographic extent of the study of Pacific Northwest old growth has increased to more than 25 million acres—all of western Washington, Oregon, and northern California—using inventory grids and satellites orbiting hundreds of miles above the earth. At the regional level, the loss of an entire stand of trees to fire or logging is not very noticeable.

Applying the Scientific Method to Study Old Growth

The application of the scientific method to the science of old growth also is challenged by the long timeframes and diversity of environments and forest types. Some determined scientists have begun to investigate how old-growth logs decay through experiments that are designed to run for 200 years. In general, however, centuries-long experiments are not feasible, not only for reasons of funding and institutional memory but also because the very subject of the study—a whole forest or study plot—may be destroyed by fire, wind, or other forces. Consequently, scientists are forced to use other less-powerful scientific methods, such as manipulative experiments that run a few years or decades, retrospective studies of ecological phenomena created by past natural or human disturbances, and computer simulations, which allow scientists to study how forests *might* change over long times or large areas. These methodological limitations do not make the science of old growth less of a science—geology and astronomy, for example, have done very well without being able to use controlled manipulative experiments.

The science of old growth has advanced by focusing on structure and process—elements that can be readily measured—rather than on questions of absence of human influence. The difficult reality is that the only places where scientists can learn about the structure and process of old forests is where the absence of human activity has allowed some populations of trees to survive for centuries—so called *virgin* or *near-virgin* forests. At a grand scale these are the scientific *controls*—places against which to compare the changes that result from the unplanned "experiments" of human activities. In fact, the Society of American Foresters recognized this value as far back as 1947 when it began to locate and protect examples of "virgin or old-growth" forests around the United States for forest research purposes (Anonymous, 1959). The need for controls is evident when one considers how intensively managed forests can be treated, not to mention how forests can be converted to residential or commercial uses. It is, however, difficult to find or define a "control" in many landscapes where human activities directly or indirectly influence the development of forests—e.g., forests with a history of fire suppression.

Theoretical Basis of Old-Growth Science

Our scientific thinking about how and why forests changes as they mature has evolved considerably over the past 100 years. In the early 1900s Frederic Clements, an American plant ecologist, proposed a model (Barnes et al., 1998) for vegetation succession in which vegetation changed in response to disturbance, migration, and establishment. Over time, he reasoned, these processes led to a stable climax forest, whose character was controlled entirely by regional climate. This forest allegedly behaved as if it were a superorganism, seeking to maintain equilibrium. Clements's theories have since been discredited and replaced by theories based on the observations that many factors control forest composition, that vegetation change is not so predictable—disturbance and climate change can intervene at any time—and that plant communities are really a loose assemblage of species that may influence their neighbors but are basically behaving individually, not as a group.

Recent Scientific Perspectives

In recent years, our scientific concepts and models of forest structure and dynamics have changed to recognize the complexity of ways forests change and the diversity of old growth. *Forests are like square pegs, and our simple models are like round holes—we have to work hard to make them fit.* In wetter areas of the Pacific Northwest, forest structure changes in a semipredictable way with stand age, as long as a stand can grow for centuries with only minor or moderate disturbances (fig. 3.2). The changes with age are gradual, but there also is much variability as a result of variation in disturbance history and site productivity. Large old live and dead trees can be found in patches in very young forests, and very young forests can be found as patches within older forests. In some areas, young natural forests can be almost as complex as old growth. Furthermore, although the large dead wood did originate in older forests, it is not as distinctive of old-growth forests as was once thought. It can occur in even greater abundance in young forests created following wildfire and windstorm. In drier environments, stand age often does not predict forest structure because trees can be of many ages, making it difficult to say what the age of the "stand" really is. The textbook model of stand development recognizes four stages, while a newer model recognizes eight stages. These might not occur in a linear sequence, and some stages might be skipped (table 3.1).

The opposing forces of patchy disturbances and decay versus biological growth further blur the boundaries of forest types. For example, although many old-growth stands in the western Cascades of Oregon established as young forests following infrequent, high-severity fires, many of these stands experienced low- to moderate-severity fires in the past 200 years that killed some overstory trees and understory vegetation, and created dense patches of young understory hemlock and Douglas-fir trees. In contrast, in

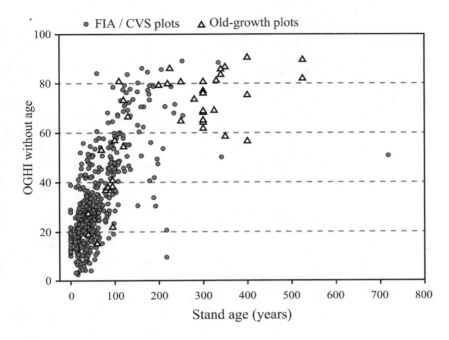

FIGURE 3.2. Old-Growth Habitat Index (OGHI) (structure-based, age not included) plotted against stand age for inventory (FIA/CVS plots) and old-growth research plots in the Oregon Coast Range.

the eastern Cascades in dry old-growth ponderosa pine and mixed conifer types, patchy surface fires burned frequently—every ten to forty years—creating relatively open understories (see chapter 1, fig. 1.2) and scattered patches of regeneration. In these forest types insects and diseases that kill old trees also have added to the structural complexity.

The possibilities of stand development really become apparent at the landscape level (fig. 3.3). The idealized developmental pathway typically presented a stand-level view in which an old-growth forest develops over many centuries with relatively little disturbance and with an orderly progression of stages, but this is only one of many pathways a forest can take, and in many regions it may not be the most common one. For example, the repeated occurrence of low- to moderate-severity fires in old forests can either set back succession, if early successional species regenerate, or advance it, if shade-tolerant species regenerate. Older forests subject to this type of disturbance regime are a complex mosaic of young, mature, old, and very old trees.

TABLE 3.1. Examples of different Douglas-fir stand development stage classifications in relation to stand age.

Typical stage age (years)	Four-stage model (e.g., Oliver and Larson, 1990)	Eight-stage model (Franklin et al., 2002)
		Disturbance and legacy creation
0		
	Stand initiation	Cohort establishment
20		
		Canopy closure
30	Stem exclusion	
		Biomass accumulation/ competitive exclusion
80	Understory re-initiation	
		Maturation
150	Old-growth	
		Vertical diversification
300		
		Horizontal diversification
800		
		Pioneer cohort loss
1,200		

The danger we face in putting forest development into just a handful of types or boxes is that we will confuse our simple models of reality with reality itself. The focus on old forest also can blind us to the importance of other stages, such as open canopy types with lots of shrubs and snags that are important to many species, including even the northern spotted owl in the southern part of its range (see chapter 4).

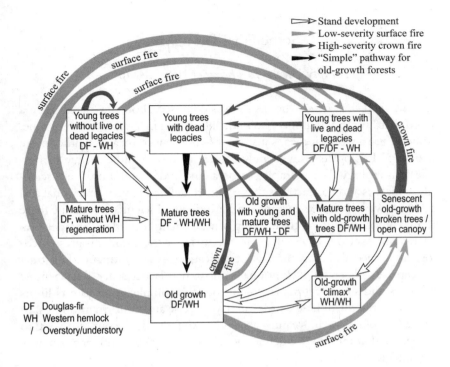

FIGURE 3.3. Conceptual model of multiple pathways of stand development at landscape level for the western Cascade Range in Oregon. (Rapp and Spies, 2003)

If you build it they will come, but they may take their time. The northern spotted owl occurs more commonly in older forests because of the presence of broken tree tops for nesting and multiple canopy layers for roosting and foraging. But not all old forest species are present in older forests because of a particular habitat structure. For example, one canopy lichen, *Lobaria oregana,* a slow-dispersing species, occurs much more frequently in old-growth forests than in young growth, but it is relatively broad in its habitat tolerance. This lichen can be transplanted to much younger forests and survive and grow very well (Sillett and McCune, 1998). Lobaria is more abundant in older forests probably because it disperses slowly from existing sites, and it can take centuries to recolonize and accumulate on a site once its populations have been destroyed by fire or logging. This finding suggests that other "specialists" of old growth may be present simply because of the amount of time that has passed following severe disturbance.

Some Final Thoughts

Humpty Dumpty's Theory of Word Meaning

In *Alice in Wonderland,* when Alice was confused about the meaning of a word that Humpty Dumpty had said, "he replied in a rather scornful tone, 'When I use a word, it means just what I choose it to mean—neither more nor less.'" Ecologists are generally comfortable with rather vague or flexible terminology. For example, "ecosystem" and "community" are scientific concepts and abstractions that often are defined in very situational and arbitrary ways. It is not uncommon for individual scientists to define these terms somewhat differently, creating private scientific models (Shrader-Frechette and McCoy, 1993) that can be barriers to scientific progress. Ecological science also runs the risk of reification—the treatment of abstract concepts as concrete. Definitional consistency is necessary for scientific progress and successful management, but the complexity of forests limits the degree to which these words can have precise and uniform meanings. In the case of old growth, we must always be careful to define our terms and to recognize that the way we define them can have a major influence on how we think about them.

The Goldilocks Problem

The problem of complexity in our scientific models of forests is like Goldilocks's problem of getting the porridge "just right" in the children's story *The Three Bears.* If we make them too complex, few people will understand the models, and managers will not use them. If we make them too simple, we miss important features of the ecology of these forests. The old-growth icon has served scientists and the public as a symbol of forest complexity, but it is a symbol of only one part of it. The unquestioned political success of the icon can also be a trap if researchers and managers do not look at the entire spectrum of forest development.

Old forests are inextricably intertwined in space and time in a continuum of forest development, just as young, mature, and mixed-age forests are. Focusing on only one part of the continuum is like trying to understand light by examining only one color or wavelength, or like trying to understand a river by looking only at the deep, quiet pools and ignoring the rapids. Forest ecology is moving toward becoming a science of complexity, but it is not there yet. Many ecologists have come to realize that it is crucial to avoid ideas that become limiting notions in our quest to better understand our world, even though this approach renders the pursuit of new knowledge a significantly more difficult enterprise. Complexity does not yield easily to simplified thinking.

LITERATURE CITED

Anonymous. 1959. Forest reserve program. *Science* 130(3370):258.

Barnes, B. V., D. R. Zak, S. R. Denton, and S. H. Spurr. 1998. *Forest ecology,* 4th ed. New York: Wiley.

Binkley, D. 2004. A hypothesis about the interaction of tree dominance and stand production through stand development. *Forest Ecology and Management* 190:265–71.

Busby, P. E. 2002. *Preserving "old growth": Efforts to salvage trees and terminology in the Pacific Northwest.* Cambridge, MA: Harvard University.

Clark, D. B. 1996. Abolishing virginity. *Journal of Tropical Ecology* 12:735–39.

Franklin, J. F., K. J. Cromack, Jr., W. Denison, A. McKee, C. Maser, J. Sedell, F. Swanson, and G. Juday. 1981. *Ecological characteristics of old-growth Douglas-fir forests.* General Technical Report PNW-118. Portland, OR: USDA Forest Service, Pacific Northwest Forest and Range Experiment Station.

Franklin, J. F., T. A. Spies, R. Van Pelt, A. B. Carey, D. A. Thornburgh, D. R. Berg, D. B. Lindenmayer, et al. 2002. Disturbances and structural development of natural forest ecosystems with silvicultural implications, using Douglas-fir forests as an example. *Forest Ecology and Management* 155(1):399–423.

Munger, T. T. 1930. Ecological aspects of the transition from old forests to new. *Science* 72(1866):327–32.

Oliver, C. D., and B. C. Larson. 1990. *Forest stand dynamics.* New York: McGraw-Hill.

Shrader-Frechette, K. S., and E. D. McCoy. 1993. *Method in ecology: Strategies for conservation.* New York: Cambridge University Press.

Sillett, S. C., and B. McCune. 1998. Survival and growth of cyanolichen transplants in Douglas-fir forest canopies. *The Bryologist* 10(1):20–31.

Thomas, J. W., tech. ed. 1979. *Wildlife habitats in managed forests: The Blue Mountains of Oregon and Washington.* Agriculture Handbook 553. Portland, OR: USDA Forest Service, Pacific Northwest Forest and Range Experiment Station.

Wimberly, M. C., T. A. Spies, C. J. Long, and C. Whitlock. 2000. Simulating historical variability in the amount of old forests in the Oregon Coast Range. *Conservation Biology* 14:167–80.

Chapter 4

Old-Growth Forest as Wildlife Habitat

BARRY R. NOON

For most humans, a walk through an old-growth forest is an awe-inspiring experience. Looking upward at the stately giants towering overhead reminds us that our lives represent but a short interval of time in the overall continuum of life. Scientists studying old-growth forests would be the first to acknowledge that, despite many new insights emerging from the study of these complex systems, there is much that we still do not understand about these forests. The complexity and variability of old-growth forests, their magnificence and spiritual presence, have inspired many to take a stand against their continued loss.

The challenge we face to conserve old-growth ecosystems, including their wildlife component, is that their uniqueness is still debated among scientists, between scientists and the public, and among diverse stakeholders. The scientists who study forest processes, structures, and species composition are constantly asked to provide guidance for appropriate old-growth forest policy. However, for the near future, our understanding will remain incomplete.

So what *do* we know about old growth? And what kinds of holes are left in the puzzle when we've listed what we know?

We know that the current areal extent of old-growth forest is significantly less than its historical distribution (Bolsinger and Waddell, 1993;

Strittholt et al., 2006) and what remains is highly fragmented (Spies et al., 1994; Nonaka and Spies, 2005). In addition, old growth is believed to have been a more persistent and predictable component of historical landscapes, although the distribution of some old-growth forest types across the landscape was dynamic due to fire and windthrow. The combined characteristics of historically broad extent and predictable occurrence suggest that many animal species have evolved in the presence of old-growth forest and adapted to exploit its resources. For those old-growth forest types characterized by stand-replacing disturbance, these adaptations would include important dispersal and colonization abilities. Wherever survival and reproductive rates allowed for self-sustaining populations, wildlife would have evolved physical attributes and behavioral responses that allowed them to persist. Thus, our understanding of the long-term co-occurrence of wildlife and old forests, coupled with the processes of natural selection and evolution, lead us to expect that many animal species would be well adapted for existence in old-growth ecosystems.

Scientists have observed that many wildlife species occur most predictably in old-growth forests (Marcot, 1997). Some of these species also may occupy younger, earlier stages of forest development, but they often occur at higher densities in landscapes where old growth is common. Also, many species utilizing younger forest may also use old growth during certain periods of their life history. For example, many common species, such as red-tailed hawks, common ravens, turkey vultures, pygmy owls, and olive-sided flycatchers, often nest in old-growth forest adjacent to open areas. Thus, the observation that many species are associated with old growth matches our evolutionary expectation. Given these relationships, it is clear that loss of old growth will be accompanied by significant population declines in some wildlife species.

High Value of Diversity to Wildlife Habitat

The most obvious feature of old-growth forest that distinguishes it from younger forest stages is its three-dimensional diversity at both local and landscape scales. This diversity is an expression of numerous physical and biological processes that accumulate organic structures, such as large trees, logs, lichens, moss, and fungus, over space and time. These processes include the local colonization, growth, and death of individual plant species, as well as broad-scale growth and mortality events caused by disturbance, disease, or insect outbreaks. These local and broad-scale events produce high spatial diversity at multiple scales, from individual trees through stands and patches to landscapes.

In addition to the presence of large trees, the presence of large amounts of accumulated dead organic matter may be the most clearly time-dependent characteristic of old-growth forests. Individual trees accumulate injury and disease events that result in structures such as cavities, broken tops, epicormic growths, and mistletoe brooms that provide nest sites for old-growth–associated species such as spotted owls, goshawks, flying squirrels, tree voles, and bats. Furthermore, mortality often is a spatially contagious process that creates large patches of dead trees that further contribute to habitat diversity. These dead trees often become nest and feeding sites for many cavity-nesting birds and flying squirrels, and when the dead trees fall to the ground they provide cover and refuge for many amphibian and small mammal species. Because these processes take time, the accumulation of features associated with mortality, injury, and disease is difficult or impossible to accelerate by management except through the judicious retention of legacies from previous forest stages.

Within an old-growth forest, the combination of vertical and horizontal diversity within and among patches, coupled with the existence of unique structural elements that accumulate over time, provides a complex habitat template for wildlife species. Increasing structural and compositional variability as forests develop through time may increase species diversity in old growth relative to many younger forest stages because of a greater number of habitat niches available for partitioning and reduced competition between similar species. The many distinct niches and habitat types within an old-growth forest effectively increase its among-habitat or ß-diversity. For habitat diversity to increase species diversity in this way, each habitat type must be sufficiently large to support a viable population. As a result, the size of old-growth reserve areas becomes exceedingly important, particularly for species with large area requirements.

Patch Size and Wildlife Habitat

It would be incorrect to view habitat selection of old-growth forest by wildlife as a threshold phenomenon. Rather, habitat selection by old-growth–associated species should be viewed as a process in which the probability of selection increases continuously as a forest ages and acquires more old-growth characteristics. Think of forests as existing along an age gradient with each patch viewed in terms of its degree of membership in the larger set of old-growth forests. The dimensions of membership are defined by the presence and amount of key structural elements such as multilayered canopies; large, old overstory trees; numerous snags; heavy accumulation of dead wood; broken-top trees; and young forest components in canopy gaps. Other "membership" dimensions include attributes of composition that are acquired with age and accumulated disturbances, such as mixtures of shade-tolerant and -intolerant species and vegetation characteristic of earlier successional stages found in canopy gaps. Without legacies from previous stands, then, early seral forest would have no membership in the set of old-growth forests. Thus, as a forest ages the likelihood increases that it will support wildlife species associated with old growth.

At broad spatial scales, forest ecosystems represent a changing mosaic of different-aged patches. Because the mosaic includes a changing distribution of forest age classes, populations of wildlife associated with old growth will be dynamic, as will the location of old growth on the landscape. The result is a shifting mosaic of compositional and structural patterns that appear to be in dynamic equilibrium only when viewed at very large temporal and spatial scales—larger than that of major disturbance events and longer than the average interval between disturbances.

Small-scale disturbances do not necessarily lead to changes in the location of old-growth patches but are important in generating within-patch diversity. The distribution of different-aged, small-scale patches embedded within the larger old-growth forest patch is an important source of habitat diversity. Thus, old-growth conservation requires management at both landscape and local scales. At the landscape scale, the extent of forest management must be sufficiently broad to accommodate the shifting mosaic of forest age classes generated by large-scale disturbances. At the local scale, individual old-growth reserves must be sufficiently large to incorporate most small-scale disturbance events that promote fine-grained habitat diversity.

Risks Associated with Reducing the Amount of Old-Growth Forest

Reducing the amount of old-growth forest will reduce the carrying capacity of the landscape for species dependent on these habitats for all, or part of, their life history. Also, many species such as small mammals and amphibians that are adapted to relatively persistent habitats often have limited dispersal capabilities. These species will be sensitive to reductions in the size or increases in the distance between remaining patches of old-growth forest. Overall, the risk of loss of old growth is more pronounced than expected solely from a reduction in area. Once lost, it will take a century or more for new habitats to develop the structural and compositional attributes needed for dependent species. In contrast, it takes relatively little time for young forest to develop. During the slow transition to old-growth conditions, species dependent on old-growth forest will remain at high risk due to small population size. Numerous extinction risks accompany small populations, and it is likely that many species could be lost or greatly reduced in abundance while awaiting a return to old-growth conditions.

Because of the inescapable time factor that distinguishes old growth from earlier forest stages, it is imperative that old growth receive special management considerations. The retention in managed stands of those structures such as large trees, snags, and downed logs that can be renewed only over long time periods, or accumulate following a sequence of disturbance events, is one way to reduce the risks of extinction during the long transition to old-growth conditions.

A Systems Perspective on Conserving Old-Growth Forests

The conservation of old-growth forest requires one to adopt a systems perspective and to acknowledge the importance of earlier successional stages. A systems perspective encompasses all stages of forest development as well as the dynamic processes that generate transitions between those stages both in space and time. Sustaining the entire successional sequence is essential to meet the needs of the full complement of wildlife species in forested ecosystems as well as to ensure the replacement of lost old growth.

A narrow focus on old growth as a terminal stage of plant community succession fails to recognize the dynamic nature of plant communities and the need for the replacement of late-successional components because of ongoing loss due to natural disturbance, advancing age, and logging. Thus, the emphasis should be on the desired landscape distribution of forest stages so that the rate of replacement of old growth meets or exceeds its rate of loss. The proportion of the landscape expected to be over a certain age can be estimated based on average transition probabilities between the stages of forest development. In this context, the absence of old growth over large areas is evidence that event(s), natural or human-caused, have truncated the successional process.

For forests subject to stand replacement disturbance events, a simplistic view of the process of forest succession considers two broad phases of development. For the first 100 years, the forest is homogeneous in terms of tree age with relatively constant spacing of stems and little variability in tree crowns. Extensive tree mortality occurs during this period via thinning from below, a result of competition for light and moisture. Over the second 100 years, the forest becomes more heterogeneous, and the dominant mortality factors are random events arising from root diseases, insect outbreaks, fire, windthrow, and drought. Forest thinning occurs as top-down as well as bottom-up mortality, creating holes in the canopy and greatly increasing vertical and horizontal diversity. All of these processes generate structural and compositional diversity that is the foundation of the niche diversification required to sustain multiple wildlife species.

For old-growth forests subject to low- to mixed-severity disturbance regimes, the aftermath of disturbance may result in a diverse array of stand conditions. Such conditions may provide the structural and compositional diversity needed to support a diverse wildlife community. However, given a mix of patch conditions, the determining factor for many wildlife species is whether the remaining patches of habitat are of sufficient size to support viable populations.

Figure 4.1 characterizes the forest as a system composed of discrete successional states—an oversimplification because forest change is a continuous process. However, representing successional change as discrete phases matches how many land management agencies think about the forested landscape. The system includes a range of states undergoing a gradual (but sometimes episodic) transition to other stages with self-renewal occurring during the old-growth phase.

Forests are in a constant state of flux with old growth as an unstable endpoint. The endpoint will remain dynamically stable only so long as the old-growth forest remains free from stand-replacing disturbance. Following large-scale disturbance, resident wildlife must be able to relocate and disperse to remaining patches of old growth. Dispersal success will be low unless the surrounding forest matrix is conducive to movement and other old-growth patches are available and close enough to accept colonists. Thus, issues of landscape connectivity, measured in terms of the likelihood of successful movement between patches of old-growth forest, become critical to sustaining wildlife over the long term. Maintaining connectivity is a significant challenge because a landscape architecture that is connected for a Vaux's swift, for example, may not be connected for an Olympic salamander or tree vole.

Forest Ecosystem

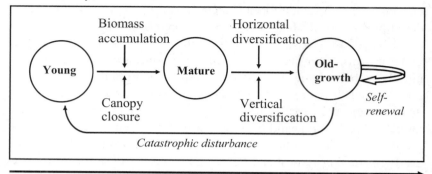

FIGURE 4.1. The forest successional process, including transitions between seral stages with some degree of self-renewal in the old-growth stage until it is interrupted by a large-scale disturbance event.

Old-Growth Forest Landscape

In the absence of catastrophic disturbance, old-growth forest can be stable in terms of area, structure, composition, and location on the landscape. If individual old-growth patches are of sufficient size and are not changed by disturbance, they can incorporate many small-scale disturbances and continue to look and function like old growth. Thus, over the short-term, a static old-growth forest reserve system appears to be an adequate conservation strategy. However, such a strategy may ultimately fail unless the system of old-growth reserves addresses the need for a dynamic balance between the rate of loss due to disturbance or logging and the rate of replacement via succession.

Critical design elements include individual reserves that are large enough to accommodate small- to moderate-scale disturbance events and a large enough set of reserves to achieve spatial redundancy at a scale that exceeds even large severe disturbance events. These design considerations were incorporated into the Northwest Forest Plan (NWFP) and its systems of late successional reserves. To date, the NWFP has stood up to large disturbances without jeopardizing species associated with old-growth forest.

Because of the dynamic nature of forested ecosystems, the location of old-growth forests will change through time, and new reserves may be required to replace those lost through natural attrition. As a consequence, zoning of the landscape into a system of reserves and nonreserves may not be a successful long-term conservation strategy. To plan for a target amount of old growth with spatial distribution changing over time requires that the forest be viewed in its entirety as a spatially dynamic system. In addition, animals must be able to move through the landscape to track the movement of old-growth patches. Therefore, the matrix in which old-growth reserves are embedded must be managed so as to maintain connectivity among reserves. For species with low mobility, management may be needed, for example, through translocations, to ensure their long-term viability.

Managing a dynamic system at a landscape scale is challenging. The static reserve design strategy for the northern spotted owl and other old-growth wildlife species in the Pacific Northwest is not certain to prove adequate in the long term. This uncertainty is not solely attributable to unpredictable severe disturbance events or climate change—the reality is that the fate of old-growth forests may depend more on future, unknowable political decisions than on environmental factors.

Source and Sink Old-Growth Forest Patches

Individual old-growth patches (or reserves) may act as source areas—where, on average, birth rates exceed death rates—for colonists to other old-growth patches or to younger forests. Several factors, local- and landscape-scale, influence whether a patch of old-growth forest acts as a *source area* (produces colonists to other patches) or as a *sink area* (acts a net receiver of colonists from source patches). Local factors include a patch's habitat quality, measured in terms of the average birth and survival rates of the focal species. If local birth and survival rates provide for a growing population, then that patch serves as a source area for the focal species. Otherwise, the patch is a sink, and its population must be maintained by immigrants from source patches. An additional local factor is the size of a patch. In general, large patches will support larger populations that are more stable across years and more likely to produce colonists for other patches.

Landscape-scale factors also influence whether a patch acts as a population source area. Even if local birth and survival rates provide for a growing population, emigrants from a patch must disperse to, and successfully colonize, neighboring patches. Successful travel between old-growth patches will be affected by how far apart they are and the nature of the habitat between patches. In general, sink patches that lie close to source patches and/or are separated by habitat that is similar in structure and composition to source patches will be more likely to receive colonists.

The important concepts of patch size and spacing, landscape context, and source and sink populations can be captured in a simple diagram (fig. 4.2). Figure 4.2 shows seven old-growth forest patches and considers their contributions to the population dynamics of a hypothetical species. Source area A is composed of three small patches, close together and strongly connected by among-patch movement, which effectively function as one large patch. Source patch B provides colonists to sink patches A, B, and C. Movement to sink B is more likely to be successful than movement to sink C because the nature of the forest matrix along this pathway is more conducive to movement. Sink A is too small to support a locally stable population and depends on immigration from sources A and B to maintain its population. Sink B is large enough to sustain a growing population, but its habitat quality is too low—average birth rates are less than death rates—to act as a source area.

Are There Any Truly Old-Growth-Dependent Wildlife Species?

Based on current information, we know of few vertebrate species that exclusively depend on old-growth forest. Dependency means that a species's life history requirements are met only within old growth or old growth is the only habitat that supports source populations. Based on past research in the Pacific Northwest (Thomas et al., 1993), candidate species for this list include several amphibian species (e.g., Oregon, Del Norte, and Pacific giant salamanders and tailed frogs), birds (e.g., northern spotted owls [box 4.1], Vaux's swift, red crossbills, several cavity-nesting birds), and mammals (e.g., several *Myotis* bat species, pine martens, Pacific fishers, northern flying squirrels, tree voles).

However, individuals who question the uniqueness of old-growth forest could point out that, although all of the above species may reach their maximum abundance in old-growth forest, they also occur in lesser numbers in younger forest. And the critics would be correct. However, the key question is whether younger-aged forests occupied by a species are, in the absence of old growth, sufficient to sustain the species over the long term. The simple observation of a species occupying and even reproducing in a habitat is not, by itself, sufficient to infer that the habitat will sustain the species over the long term. In the absence of data that link birth and

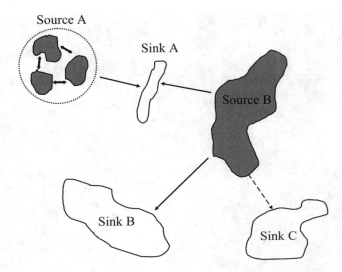

FIGURE 4.2. Distribution of source and sink old-growth forest patches and their connectivity in terms of a hypothetical old-growth–dependent species. Solid arrows indicate high connectivity. Dashed arrows indicate low connectivity due to the nature of the forest matrix.

death rates to specific habitat types, we work under the assumption that those habitats in which the species is more likely to be observed or where it reaches its greatest abundance are most important to its long-term conservation. Based on this criterion, we infer that many species appear to be "dependent" on old-growth forest.

The issue of dependence is further complicated by the observation of species that use multiple habitat types to meet their life history requirements. For example, a species may use old-growth forest to meet its nesting requirements but use forest edges or younger forest to meet its requirements for food or cover. This appears to be the case for northern spotted owls in the southern part of their range. In areas where dusky-footed woodrats are its primary prey, spotted owls exploit the edge between old-growth and young forest because woodrats are more abundant and available there. In areas where northern flying squirrels are the owl's primary prey, the association with old-growth forest is much stronger. However, even where woodrats are exploited, the owl still requires old-growth structures—large old trees with cavities—to meet its nesting requirements.

The information needed to identify those species for which old-growth forest acts as essential source habitat is unknown for most species. However,

FIGURE 4.3. Northern spotted owl. (Photo: T. Iraci)

BOX 4.1 THE NORTHERN SPOTTED OWL

Eric Forsman

As a young biologist in the early 1970s, I was convinced that if we stopped clearcutting old forests on federal lands in the Pacific Northwest, northern spotted owls (fig. 4.3) and other species that occurred there would do just fine. I am no longer so sure. I have come to realize that ecological factors other than habitat, like climate, fire, and competing species, make long-term predictions about species very iffy. This realization, which should not be a surprise to any student of Darwinian evolution, is nevertheless unsettling to most humans, who look with nostalgia at the past and with fear at the uncertain future.

The recent invasion of the Pacific Northwest by the barred owl, which is closely related to the spotted owl, is now thought to pose a new threat to the spotted owl. This has flummoxed the conservation agencies and resulted in some rather odd reasoning. For example, the U.S. Fish and Wildlife Service Spotted Owl Recovery Team recently proposed (in a draft plan that was changed after public review) to "recover" the spotted owl by conducting a large-scale barred owl removal experiment while at the same time reducing the amount of habitat currently protected for the spotted owl. A conservation plan based on shooting barred owls and at the same time further reducing the habitat for a threatened species makes little sense to a biologist.

Although they actually occur in a fairly broad range of forest types and age classes, spotted owls tend to be most common in old forests. This has led most owl researchers, myself included, to conclude that the best way to benefit the owl is to save as much as possible of the existing old forest and to grow new areas of old forest on at least some areas that were logged in the past. But the devil is in the details. For example, some recent studies suggest that, in those areas where they feed mainly on woodrats, spotted owls actually do best in landscapes that include a moderate amount of forest fragmentation—a mix of old forests and brushy clearcuts. It also is likely that, in fire-prone landscapes, areas of open forest that may not be particularly good habitat for spotted owls may play a crucial role in slowing the spread of fire into the dense old forests that are preferred habitat. These complicating factors often get ignored or over-emphasized, depending on who is doing the talking.

The spotted owl often is thought of as an "umbrella" species, whose habitat will also harbor many other species, thereby avoiding the need to develop individual management plans for hundreds of species. This assumption may work for some species but doesn't work well with others. For example, some rare species may occur only in areas outside the areas managed for owls. Because of this concern, the NWFP included a requirement that federal managers had to conduct surveys for a long laundry list of poorly known or rare species before they could do any cutting in areas outside of the old forest reserves. These surveys turned out to be expensive and time-consuming and, in combination with lawsuits challenging management activities, were one reason that the timber harvest targets in the NWFP have not been met.

The debate over how to manage the spotted owl is a classic example of why a single-minded focus on a single species can't solve a problem that is really about how to manage an entire ecosystem, including the humans in it. The old forests that remain on federal and private lands in the Pacific Northwest are products of a unique series of events that will never be repeated. So the notion that we can ever exactly replace these forests once they are cut is both fanciful and false. Once cut, these old forests will become ghosts from our past, invisible to our children's children except in picture books of loggers lying in the face cuts of giant trees. The solution then must lie in some combination of species and ecosystem approaches, which also take into account the always shifting variables of human preference. For the owl and the old-growth "problems" have as much to do with our social values as they do with our science.

given our inability to rapidly replace old growth once it is lost, prudence suggests that we devote considerable effort to old-growth conservation and invest in the necessary research until we better understand the role of these forests as refuges and sources of colonists for sustaining biodiversity across the forested landscape.

Conclusion

Old-growth forest provides unique habitat for many wildlife species for at least some component of their life history. In the absence of a sufficient amount and appropriate distribution of old-growth reserves, we can expect that some species will be lost or greatly reduced in abundance. Because of the dynamic nature of old-growth forest, the threat of loss due to stand-replacing disturbance, and the unknowns associated with global climate change, maintaining old-growth forests on future landscapes presents significant management challenges. Meeting these challenges requires viewing old growth from a landscape perspective and as one component of the larger forest ecosystem that includes all forest age classes. The concept of the distribution of different-aged forest stands as a shifting spatial mosaic is relevant in the context of rapid landscape change and global climate change. This reality requires managers to adopt a dynamic planning paradigm and to incorporate redundancy at broad spatial scales into the planning of old-growth reserves. Such a strategy is needed to maintain resilience in the face of surprise environmental events. Whether state and federal land management agencies are capable of implementing a dynamic spatial design for old-growth reserves given accelerating rates of landscape change is highly uncertain at this time. This uncertainty is attributable, in part, to incomplete scientific information and understanding but perhaps even more to a constantly shifting and therefore inadequate political and social commitment to protecting our natural heritage.

LITERATURE CITED

Bolsinger, C. L., and K. L. Waddell. 1993. *Area of old-growth forests in California, Oregon, and Washington*. Resource Bulletin PNW-RB-197. Portland, OR: USDA Forest Service, Pacific Northwest Research Station.

Marcot, B. G. 1997. Biodiversity of old forest of the West: A lesson from our elders. In *Creating a forestry for the 21st century: The science of ecosystem management*, edited by K. A. Kohm and J. F. Franklin. Washington, D.C.: Island Press.

Nonaka, E., and T. A. Spies. 2005. Historical range of variability in landscape structure: A simulation study in Oregon, USA. *Ecological Applications* 15(5): 1727–46.

Spies, T. A., W. J. Ripple, and G. A. Bradshaw. 1994. Dynamics and pattern of a managed coniferous forest landscape in Oregon. *Ecological Applications* 4:555–68.

Strittholt, J. R., D. A. DellaSala, and H. Jiang. 2006. Status of mature and old-growth forest in the Pacific Northwest. *Conservation Biology* 20:363–74.

Thomas, J. W., M. G. Raphael, R. G. Anthony, E. D. Forsman, A. G. Gunderson, R. S. Holthausen, B. G. Marcot, G. H. Reeves, J. R. Sidell, and D. M. Solis. 1993. *Viability assessments and management considerations for species associated with late-successional and old-growth forests of the Pacific Northwest*. Portland, OR: USDA Forest Service, Pacific Northwest Region.

Chapter 5

Maintaining Biodiversity in Managed Forests

ANDREW B. CAREY

Looking back across nearly three decades of debate about the values of old growth, a central question emerges for wildlife ecologists: Is advanced age necessary for a forest to support most plants and animals? The answer is no, although it certainly helps. Let us then broaden the question: What is the importance of biodiversity in forests other than old growth? The multifaceted Pacific Northwest experience in wrangling over old growth has actually generated much deep thought and considerable guidance for the future management of temperate forests of all ages.

Some eight million acres of older forests on federal lands in the Pacific Northwest have been set aside primarily to sustain threatened species such as the spotted owl, marbled murrelet, and Pacific salmon and to maintain the diversity of lesser known plants, animals, and fungi. The framers of the Endangered Species Act saw threatened species as indicators of failing ecosystem health and function; the framers of the National Forest Management Act saw biodiversity as an indicator of the multiple values people have for forests. Contemporary ecologists view biodiversity as integral to forest health, providing a package that includes resistance of ecosystems to actual disturbance, resilience in the face of disturbance, and the ability to adapt and persist through both minor change and major disturbance.

Youth and the Value of Legacies

But what of the large acreage of younger forests on federal, state, and private lands? Is biodiversity important there? Four million acres of actively managed young forest on federal land serve as the surroundings or "matrix" for the twenty million acres reserved. Perhaps another fifteen to twenty million acres on various ownerships are or could be managed. What role could they play in recovering threatened species and in providing the economic goods and ecological services that are needed for environmental, economic, and social sustainability?

In the 1980s, a consortium of federal, state, and university ecologists, more than 125 in number, sought to address this question and others by comparing the plants, fungi, and wildlife in naturally young (forty to eighty years old), mature (100 to 200 years old), and old (more than 200 years old) forests in western Washington, Oregon, and northern California. They found no species *confined* to old growth. The northern spotted owl did appear to require large amounts of old forest, even though it used younger natural forests and some second-growth (previously logged) forests; marbled murrelets and Vaux's swifts appeared to need large, old trees for nest sites; and salmon needed the cool, clear water and well-aerated gravel streambeds often found in old forests but also found in well-managed second-growth forests. Many species were more abundant in old growth than in younger forests, and old growth often had more diverse *biotic communities* (groups of species in an area that may interact with one another) than younger forests. In Oregon, for example, woodpeckers were ten times more abundant in old growth than in young forests, and the various species of woodpeckers were more likely to be found together in old-growth forests than in young ones, where often one or two species might be absent. Few differences were seen in migratory bird communities in natural forests of different ages; total numbers were somewhat lower in younger forests. Certain salamanders that live only in large decaying logs—for example, the Oregon slender salamander, which is found only in western Oregon—were much more likely to be found in old forests than in young forests, as were certain types of bats. Most natural forests had small-mammal communities with all the expected species, but in some young forests, certain species were more abundant, and others less abundant, than in old forests.

Still, the scientists concluded that the naturally young forests often had so many legacies from the preceding old forest that the habitat requirements of most species were met there. These legacies included large live trees, large dead trees standing and on the ground, deep organic soils, and diverse species of plants and fungi, including multiple species of trees. Thus a great deal of overlap exists among young, mature, and old-growth forests for different plant and animal communities. Many (but not all) young, natural forests, then, had much of the biocomplexity of older forests due to the legacies they received, the variability in density of trees, distribution on the landscape, species of trees that regenerated, and the ability of various species of plants and animals to recolonize the recovering forest.

Second-Growth Diversity

What about diversity in second-growth forests—those originating after timber harvests? Scientists speculated that as many as 400 species of less-studied organisms, especially many fungi, lichens, mosses, liverworts, slugs, and snails, might not persevere in second growth. For those species that were studied, the answer, of course, depended on how those second-growth forests were managed. Second-growth forests are diverse in character. They range from third- or fourth-generation intensively managed plantations ten to forty years old to forests selectively logged more than 100 years ago and left with as many legacies as forests arising after natural catastrophes. The former often lack many native species, such as some lichens, salamanders, and small mammals, and support others in low numbers, such as woodpeckers. The latter often support even spotted owls and many of the 400 lesser known species of snails, fungi, and plants mentioned above. Naturally rare species are often assumed to depend on some "climax" condition such as old growth but in fact may be limited by soil, coarse woody debris, microclimate, disturbance, or other factors. Most rare species seem to be abundant somewhere, in some narrowly defined niche. Some species, however, especially *relictual species* (those surviving in unique refuges after a major geological event such as glaciation), are truly rare and easily threatened by natural or human-induced change.

What About Plantations?

Intensively managed plantations focused on wood production are intentionally managed to be a harsh environment for many native species. The goal is often to concentrate all the resources of the site (sunlight, air, water, and nutrients) into one selected tree species. The degree of management intensity varies with ownership, with federal forests least intensively simplified, state forests intermediate in intensity of management, and industrial forests most intensively managed. Intensive management often entails clearcutting all stems to two inches in diameter (although state regulations may require retaining a specified number of trees) followed by piling and burning the logging debris and whatever live vegetation might be present, and sometimes even leveling the soil. Industrial foresters often apply chemicals to prevent the germination of plant seeds, plant genetically selected tree seedlings in high densities, and apply herbicides to eliminate emerging vegetation that might compete with those seedlings. Many foresters maintain dense stands of planted trees, using precommercial thinning to promote growth and forestall mortality due to intense competition but retaining enough "crop" trees to exclude other plants. In intensively managed plantations, the retained trees may be pruned to promote knot-free wood and produce a foliage-free gap from the base of the canopy to the litter-covered forest floor with, albeit unintentionally, not even a place for a bird or squirrel to sit. The most obvious character of such plantations is their uniformity and simplicity: evenly spaced trees of a single species with little or no understory. These plantations can be extremely efficient in turning sunlight and carbon dioxide in the air into wood in the short term.

However, very few other species can compete with the trees in such an intensively simplified, homogeneous environment. The epitome of this kind of management in the Pacific Northwest is the hybrid poplar plantation in which exotic plants (introduced weeds that colonize the bare soil under the trees) and mammals (such as old world mice and rats adapted to human activities) are more common than native plants and animals. Unfortunately for many landowners who were lured by the promise of quick financial return on investment a decade or so ago, the market for these trees plummeted recently, and they are left with ecologically and economically undesirable plantations on their land. Even using native species, *monocultures* (single-species plantations) create a level of crowding that violates basic principles of *epidemiology* (the science of the spread of diseases) and has resulted in health problems due to root rots, Swiss needle cast, beetle outbreaks, and other diseases as well as increased susceptibility to damage by wind, ice, and snow.

Plantations do exist that have been managed for multiple species or have naturally recruited multiple species despite efforts to eliminate them, and these forests have a greater capacity to support a wide array of native species. The more legacies retained from the previous stand, the more species of trees (conifer and hardwood) regenerating, and the more variability in tree density, the greater the variety of native species supported.

Understanding *Niche Differentiation*

Variety and variability beget biological diversity—hence our association of biodiversity with old-growth forests. Why is this so? A forest can be viewed as having multiple dimensions. The most obvious are the length and width of its boundaries and the height of its trees; multiply these together and the product is the volume of space occupied by the forest. Older forests occupy more space per area because the trees are taller and therefore possess a greater overall capacity to support life. Other space-per-area dimensions include the depth (shallow to deep) of crevices in the bark and the size of branches of the trees. Both of these attributes vary with size and species of tree. Different depths of crevices support different species of insects, some of which, for example, support spiders that prey upon them and are preyed upon by the brown creeper, a bird that feeds on spiders in bark crevices. Thus, deep crevices are an important niche dimension for the creeper. Another dimension is that of the size of cracks in and cavities behind plates of bark that provide nest and roost sites for the creeper.

Different sizes of branches provide foraging sites for different sizes of woodpeckers; also important to woodpecker foraging is the degree and kind of decay in boles and branches of trees. Furthermore, woodpeckers excavate cavities for nesting and roosting in decayed trees, with the smallest woodpecker requiring the most decay and the largest woodpecker requiring both large trees and decay only in the heart of the tree. Woodpeckers, then, partition, or differentially use, the multidimensional space provided by different species and sizes of trees, different sizes of tree branches, and different degrees of decay. The multidimensional space actually occupied by a species, in the presence of somewhat similar species, is its *niche*. But, as can be imagined, a complex forest with substantial variation in several dimensions provides a variety of niches potentially available to a large number of species, even before the species are present and can interact to partition the resources available.

Thus, an old forest with a wide variety of tree species, tree and branch sizes, partially dead trees, standing dead trees, and fallen trees allows all the members of the woodpecker community, which differ in body and bill size, to coexist in abundance without undue competition between them.

And so it goes for biotic community after biotic community in forests of all ages: The forest is multidimensional, and the variety within each dimension interacts with other dimensions to provide a diversity of niches available for plants, fungi, and animals to occupy. This property is the basis for maintaining high biological diversity (and its associated resilience and forest health) in both natural and managed forests.

Untangling Biocomplexity Through Food Webs

Over a twenty-year program of research, I have purposefully contrasted naturally young, managed young, and old-growth forests to determine to what degree biodiversity could be retained, maintained, or promoted in second growth. I have also examined how that might best be done while providing other values such as air, water, wood, recreation, wildlife, and fish. A major research challenge was that much of biodiversity is unknown to us; most species have yet to be described. The vast majority of species are small, even invisible to our eyes: parasites, fungi, bacteria, and viruses, especially in the soil food web but also in the gut of animals. So I had to choose some surrogates of overall biodiversity based on ecosystem function.

First, I identified a food web that was symbolic of Pacific Northwest old growth and that represented key ecological functions. This is called a *keystone complex*, and the example I identified was northern spotted owls–northern flying squirrels–ectomycorrhizal fungi–Douglas-fir (fig. 5.1). The spotted owl is the flagship species for old growth; to it can be added goshawks, other owls, ermine, long-tailed weasels, marten, and fisher. The flying squirrel is the primary prey of the owl and marten. To the flying squirrel we add the Douglas's squirrel, a favorite prey of goshawks and fisher, and Townsend's chipmunks, favorite prey of smaller weasels. Hypogeous ectomycorrhizal fungi surround and penetrate the fine roots of Douglas-fir and produce spores (the fungal equivalent of a seed) packed in an underground ball called a *sporocarp*; the ball emits odors that attracts all three squirrels. These sporocarps are the primary food of the flying squirrel and are important food for the other squirrels. The squirrels digest the outer sporocarp but not the spores; they then disperse the spores and associated microorganisms throughout the forest in their feces. Meanwhile, the fungi enhance the ability of Douglas-fir to absorb water and nutrients from the soil and receive carbohydrates (produced by photosynthesis in the fir's needles) in return. They also may be important in protecting the roots from disease.

FIGURE 5.1. A keystone complex representing a complex food web: northern spotted owl, northern flying squirrel, ectomycorrhizal fungal spore, and Douglas-fir.

It is thought that a diversity of fungi is important to serve a diversity of trees under widely varying soil, weather, and climate conditions. The fungi move photosynthetic carbohydrates from trees to the portion of the soil around the roots and fungi that is packed with biologically active microorganisms, providing support for that vast array of microbes and other soil organisms that make the soil rich and productive. Above ground, the food web expands to include the seeds, fruits, and mushrooms that the squirrels eat, as well as various trees and shrubs that produce seeds and fruits and also support the fungi.

This food web complex provides a framework that is useful in evaluating forest ecosystem response to forest management. The spotted owl and other predators, however, respond to collections of ecosystems (called *stands* by foresters) at the landscape scale, so it is difficult to directly measure their response to stand-level forest management. But their prey can be used to evaluate ecosystem management at small scales. Thus, I first measured abundances of the three squirrels—the northern flying squirrel, the Douglas' squirrel, and Townsend's chipmunk—in second growth with various management histories and compared those abundances to the simultaneously high abundances of all three species in complex old forest. The combined biomass of the three species reflects available food resources and therefore is a measure of ecological productivity. The diets of the three species overlap, but the flying squirrel is a truffle specialist, the Douglas's squirrel is a conifer seed specialist, and Townsend's chipmunk is a fruit generalist, feeding on conifer seeds, seeds and nuts of deciduous trees, berries of shrubs, and truffles but often hibernating during food shortages in the winter. Thus, these species represent the reproductive fruits of the forest ecosystem and the capacity of the ecosystem to support diverse predators.

Second, I was concerned with forest-floor function, because the soil, its decaying organic matter, and its live inhabitants are fundamental to forest sustainability. Operating heavy equipment, killing trees, and altering microclimate by removing trees affects forest-floor function. A basic feature of Pacific Northwest forest soils is dominance of biological activity by fungi, particularly ectomycorrhizal fungi. My research teams measured activity of belowground fungi with several variables. Food webs in the forest floor support a diversity of small mammals; thus I used the number of small mammal species and their abundances as a measure of forest-floor food web complexity. This community is particularly diverse in Pacific Northwest forests compared to the rest of the world (Carey, 2003). Because so many animals respond to composition of the plant community on the forest floor, I also measured the abundance and diversity of plants.

Finally, resident birds play important roles in regulating forest insect populations by preying upon insects. These birds are particularly sensitive to the simplifying effects of timber management. Thus, I measured their diversity and abundance in winter.

Forest Management and Biodiversity

By now, the results of my investigations should be no surprise. Simple forests managed for timber—planted and thinned but left to grow for eighty years or more—did develop diverse understories. These understories supported abundant small mammals but were dominated by a few fern and shrub species and had some mammal species missing (for example, Keen's mouse), some rare (for example, the northern flying squirrel and Douglas's squirrel), and some very abundant (deer mice, creeping voles, and Townsend's chipmunk). Root rot and beetle attack were becoming problematic. Spring (migratory) birds were relatively abundant, but because trees were of the same size and unhealthy trees had been intentionally removed during thinnings, there were few woodpeckers.

Forests that had been harvested with some legacies retained and allowed to regenerate and grow without further management were dense, crowded, had obvious root rot problems, poor understory development, incomplete small mammal communities, moderate numbers of flying squirrels, few chipmunks or Douglas's squirrels, and reduced numbers of birds.

I arranged for both types of forest to be thinned to produce variability in canopy density. The resulting increases in diversity of plants, below-ground fungi, mushrooms, small mammals, birds, and squirrels were dramatic. Geographically extensive surveys of second-growth forests showed that the biocomplexity provided by multiple niche dimensions directly affected both the diversity and total abundance of small mammals (fig. 5.2).

This result illustrates that individual species do not respond to environmental change simply on the basis of the abundance of one or two of their important habitat elements. Instead, a synergy from multiple niche dimensions allows the coexistence of multiple species in high numbers. Without these dimensions, species might drop out, but we can't entirely predict which ones will do so first.

Active Intentional Management

The studies discussed here (summarized in Carey and Peeler, 1995; Carey et al., 1992, 1999a; Carey, 2003) provided a basis for a new type of management—*active intentional management* (AIM)—for biodiversity and other forest values. Carey et al. (1999b) modeled differences in numerous variables among AIM, no management, and management to maximize financial returns from timber. We simulated these three management scenarios using data from a real landscape: one that was entirely second growth, about fifty years old, and ecologically degraded.

The results were striking. Undertaking no management activities actually delayed recovery of the landscape. Timber management further degraded the landscape and put dozens of species at risk. AIM, as it was

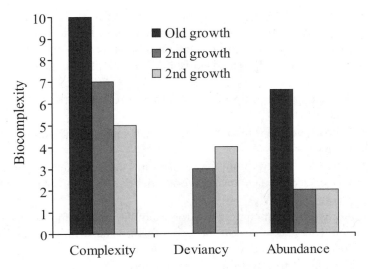

FIGURE 5.2. The relationship between biocomplexity, on a scale of one to ten, in one broad-scale sample of old growth and two broad-scale samples of managed forests on the Olympic Peninsula. The abundance of small mammals is based on captures per 100 trap nights, and the deviancy is the degree to which small mammal communities in young forest deviate (species lost, species decline in relative abundance, and species increase in relative abundance) from those in old growth.

designed to do, produced a landscape with high biodiversity: spotted owls, all wildlife whose habitat requirements could be modeled (130 species), deer and elk equivalent to the clearcut and short rotation timber management, complete small mammal communities, indicating complete soil food webs, and high numbers of the squirrels, indicating high ecological productivity. AIM also produced wood volumes nearly equal to timber management, a greater variety of wood products, greater sustainable revenues from timber, more forest-based employment, and higher tax revenues, all good indicators of sustainability.

What was lost? Six to fifteen percent of short-term economic gains, depending on degree of riparian protection. How was it done? AIM, in a nutshell, includes

- Watershed analysis that identifies areas of unstable slopes, stream sides that need protection, and areas where roads should and should not be constructed;

- Retaining legacies at every step of management, especially when capitalizing on headwater seeps and streams, soil deposited in gullies that erodes if disturbed, and unstable slopes (areas prone to shallow-rapid or deep-seated landslides) that should not be disturbed in any case;
- Managing for multiple species of trees throughout the management cycle;
- Precommercial thinning to maintain diversity and promote growth of trees;
- Variable-density thinnings to harvest wood products, induce heterogeneity, manage coarse woody debris, and promote growth of trees;
- Long management cycles—for example, alternating seventy-year and 130-year cycles on industrial forest versus the more common forty-year cycles that have severe ramifications in terms of cumulative effects and landscape character; and
- Variable retention harvest systems that maintain legacies, minimize unnecessary site preparation, and maintain soil food webs.

The Washington Department of Natural Resources has adopted the AIM approach on a large part of its ownership. The Olympic National Forest and some other national forests in the Pacific Northwest also have, along with some land-managing conservation groups and some private timber management investment companies.

Managing Forests at All Ages

Forests of all ages can be managed for biodiversity. Forest management cycles of less than 200 years won't produce all the biocomplexity of old growth, but AIM can contribute substantially to general sustainability. Even though the simulations above were later repeated on different land bases by various university, state, and private groups and the qualitative results were confirmed, it must be remembered that all simulations depend upon assumptions that may be invalidated as the biotic, social, and economic environment changes.

Furthermore, we must expect ecological surprises such as volcanic eruptions, climate change, major droughts and fires, insect and disease outbreaks, introductions of pests and diseases, invasive species, and other unknowns that will affect forest ecosystems. These will happen. Managing for biodiversity will help prepare forests to change, but there will be catastrophes, and the nature of change is not predictable. Socioeconomic forces can be formidable. Socioeconomic emphasis on net present value requires reimbursing landowners (or investors) in some fashion for altering their management. Public demand and willingness to pay a premium for "green" wood products helps to provide such incentives. Cheap imported wood is as functional as low-quality wood produced domestically in intensively managed plantations, but neither provides the ecological (water, air, wildlife habitat) or social (recreational and spiritual opportunities, open space) services of complex forests supporting local communities. The consumer must be willing to support the landowners producing the latter.

LITERATURE CITED

Carey, A. B. 2003. Biocomplexity and restoration of biodiversity in temperate coniferous forest: Inducing spatial heterogeneity with variable-density thinning. *Forestry* 76(2):127–36.

Carey, A. B., and K. C. Peeler. 1995. Spotted owls: Resource and space use in mosaic landscapes. *Journal of Raptor Research* 29:223–39.

Carey, A. B., S. P. Horton, and B. L. Biswell. 1992. Northern spotted owls: Influence of prey base and landscape character. *Ecological Monographs* 62:223–50.

Carey, A. B., J. Kershner, B. Biswell, and L. D. de Toledo. 1999a. Ecological scale and forest development: Squirrels, dietary fungi, and vascular plants in managed and unmanaged forests. *Wildlife Monographs* 142:1–71.

Carey, A. B., B. R. Lippke, and J. Sessions. 1999b. Intentional systems management: Managing forests for biodiversity. *Journal of Sustainable Forestry* 9(3/4):83–125.

Chapter 6

Fish and Old-Growth Forests

GORDON H. REEVES AND PETER A. BISSON

To the public, resource managers, and many scientists, good fish habitat and strong and diverse wild salmon populations are synonymous with *old-growth forests* in the Pacific Northwest, defined here as forests dominated by trees more than 200 years old. There is little argument that large wood produced by old-growth forests is an important habitat component in all stream sizes, from the steepest headwater channels to the largest floodplain rivers. However, old-growth forests represent a limited set of the overall range of conditions observed in pristine landscapes, and relying on streams in old-growth forests to provide the best habitat for fish (fig. 6.1) may not produce expected conservation benefits.

Nonetheless, selected attributes of streams (e.g., number of pools or pieces of large wood) running through old-growth forests are used to develop reference "standards" by which the condition of streams in watersheds with management activities is assessed. These standards may also be used to establish habitat goals for restoration efforts or to provide a framework for evaluating habitat quality. Streams in old-growth forests have often been used as "controls" in studies that examine the effect of land management activities on fish and fish habitat, such as changes in the number of landslides or in peak flows.

In a similar vein, conservation plans for fish and other aquatic

FIGURE 6.1. Editorial cartoon by Jack Ohmann. (© 2006, *The Oregonian*. All rights reserved. Reprinted with permission.)

organisms in the Pacific Northwest tend to focus on watersheds with larger intact areas of old-growth forests to anchor population recovery. As part of the Aquatic Conservation Strategy of the Northwest Forest Plan in 1993, abundance of old-growth forest was a primary factor in identifying "key" watersheds to protect high-quality habitat and to prioritize restoration of degraded streams possessing at-risk salmon and trout. Watersheds where maintaining fish populations and their habitat received special emphasis under a revised land management plan for the Tongass National Forest in southeast Alaska also contained abundant old growth (fig. 6.2).

The implication of these management policies is that streams in old-growth forests provide something akin to perfect salmon habitat. Although we believe many attributes of old-growth forests are important for maintaining healthy fish populations, the past forty years of scientific studies have demonstrated that simply maintaining old growth, by itself, is insufficient to support the full spectrum of ecological processes needed to sustain productive aquatic ecosystems. The problem is that old-growth conditions were initially believed to be stable and relatively free from natural disturbance. Based on today's knowledge of dynamic watershed processes, however, old-growth forests that are protected from all disturbances, natural or

FIGURE 6.2. An old-growth forested stream in the Tongass National Forest of southeast Alaska, illustrating cold, clear water associated with a dense forest canopy. (Photo: P. Bisson)

anthropogenic, do not necessarily provide conditions that favor high levels of production of many salmon and trout species.

The purpose of this chapter is to examine relationships among fish, fish habitat, and old-growth forests. We review the history of research that led to the perception that salmon and trout require old growth and ask if this perception has remained valid in light of our developing understanding of aquatic ecosystems and watershed processes. We stress that our intent is not to diminish the need to conserve old-growth forests; rather, we believe it may be more accurate to say that old growth is necessary but not sufficient for maintaining salmonid habitat.

Early Research

For decades it was assumed, and then documented, that land management activities could result in damage to streamside (riparian) forests. However, the basis for the assumed dependency of fish on old-growth forests was less obvious. In the late 1970s and the early 1980s, numerous studies

(Franklin et al., 1981) conducted at the H. J. Andrews Experimental Forest in Oregon's western Cascades documented the ecological properties and uniqueness of old-growth forests in the Pacific Northwest. Prior to and even during this time, old-growth forests were considered decadent forests and regarded as biological deserts by many land managers and policymakers. There were calls by federal officials to dramatically reduce the amount of old growth to establish commercial forest harvest rotations of sixty to eighty years. Research at the H. J. Andrews Forest found that old-growth forests were far from decadent and in fact contained a wealth of biological diversity. Furthermore, many of the plants and animals appeared to be strongly reliant on conditions associated with old-growth stands.

Studies on old-growth forests included aquatic as well as terrestrial ecology, and new insights into many aspects of aquatic ecology in the Pacific Northwest (Franklin et al., 1981; Sedell and Swanson, 1984) and Alaska (Murphy et al., 1984) resulted from this work. One principal finding of the studies from the aquatic perspective, and one that has had a major impact on management and policy, was documentation of the importance of large wood in streams to fish habitat. Large wood from trees of the size characteristic of old-growth forests created pools, stored and stabilized gravels, and provided cover for fish and a substrate for aquatic invertebrates, important foods for fish. The recognition that large wood was important contradicted the previously held notion that large wood (especially in log jams) often impeded fish movement and constituted a habitat problem. Previously, management agencies actively removed wood from streams throughout the western United States and Alaska to improve streams for fish. Research on the ecological importance of large wood became the impetus for the reintroduction of wood into streams as part of restoration projects and also for new policies aimed at managing riparian areas as sources of future wood for streams. Because much of the research was done in streams in old-growth forests, it was generally assumed by many managers and scientists that streams in old-growth forests held the best habitat for fish.

Is the Perceived Link Between Old Growth and Fish Habitat Valid?

The assumption that fish and old-growth forests are tightly linked persists today, but results from recent studies show that the relationship between fish and forests is much more complex and deserves continued examination.

Until recently, much of the management and research focus for fish ecology and conservation has been carried out at relatively small spatial scales, such as individual habitat types and stream reaches. Habitat requirements of individual species or local species diversity are of primary interest at these scales. But management policies and recovery plans for fish populations listed under the Endangered Species Act have been prompted by better scientific understanding of larger spatial and temporal scales (Benda et al., 1998; Reeves et al., 1995). As a result, the spatial scale of management attention has moved somewhat from the scale of individual streams to multiple watersheds.

The larger spatial extent requires that watershed processes be considered in the context of time scales of decades to centuries. Long-term changes have not previously been a major consideration when describing the productivity of aquatic ecosystems. Until very recently, streams were often assumed to be relatively stable through time and likely to recover relatively quickly if they were disturbed by natural events. Unmanaged forests, in contrast, have for some time been understood as very dynamic, with stand ages comprising a series of patches whose location varies over the landscape through time.

The life histories of many salmonids suggest that they are highly adapted to disturbance-prone environments (Reeves et al., 1995; Wondzell et al., 2007). There is a tendency for a relatively small fraction of individuals in populations of anadromous salmon and trout to stray from their stream of origin. Also, some juveniles may emigrate from their natal stream to other streams during the freshwater phase of their life cycle. In resident trout and char, some fish may undertake extensive movements throughout the drainage network, while others remain within a very limited part of a watershed. The migrants can occupy new habitats that become available following disturbances. Additionally, salmonids have a relatively large number of eggs for their size, compared to other fish that spawn in the gravel, and they may construct *redds* (spawning areas) in multiple locations during a given spawning season. The high fecundity facilitates lifecycle risk-spreading by allowing offspring to rapidly populate new areas and ensure that at least some survive.

There is an emerging understanding that streams may be very dynamic in space and time and that they can experience a wide range of conditions, just as do the terrestrial ecosystems in which they are embedded. Indeed, the association of salmon and old growth cannot be considered without taking into account extensive variation within watersheds and across regions. Several examples are used here to illustrate this important point.

Oregon Coast Range

The Oregon Coast Range experienced a history of large wildfires occurring on average every 150–350 years (Reeves et al., 1995). Landslides often followed these fires, inundating stream channels with thick layers of coarse and fine sediment. Large amounts of wood and boulders likely also entered the channels, but much of it was buried (fig. 6.3a). The primary fish habitats in the summer were pools that became isolated from each other because much of the flow passed through rather than over porous gravel. Fish numbers were relatively low, dominated primarily by coho salmon. These conditions may have persisted for as long as eighty to 100 years in heavily burned watersheds.

During the early part of the second century after a large wildfire, habitat for fish became diverse and complex. The amount of gravel decreased as it was eroded or transported downstream, exposing previously buried wood (fig. 6.3b) and sculpting pools. Additionally, as the surrounding forest recovered, wood was recruited from the adjacent riparian zone. During this phase of recovery, fish abundance (of all species) and habitat quality were high. Preliminary estimates suggest that these conditions probably existed over thirty to sixty percent of the forested landscape of the Oregon Coast Range before European settlement but shifted over time.

Habitat conditions for fish likely declined as the old-growth forest developed. A dense, shady multilayered canopy inhibited algal and invertebrate production. The amount of large wood in the channel increased with increased input from the aging forest. However, the loss of gravel exceeded the input rate of new gravel, and as a result the streambeds likely contained large expanses of bedrock (fig. 6.3c), in which pools were infrequent and of low habitat quality. Fish numbers were low, again dominated by coho salmon.

Interior Pacific Northwest

Streams in the more arid regions of the Pacific Northwest, particularly those flowing through meadows, may also pass through a range of conditions in response to periodic wildfires (Rieman and McIntyre, 1995). Following wildfires, intense thunderstorms can result in gully erosion that delivers sediment to streams in the valley. Deposition of sediment in the channel helps maintain the water table at a high level and provides the setting for

a. 80–100 years after wildfire

b. 120–140 years after wildfire

c. More than 200 years after wildfire

FIGURE 6.3. Potential range of conditions that streams in the Oregon Coast Range historically experienced. (Photos: G. H. Reeves)

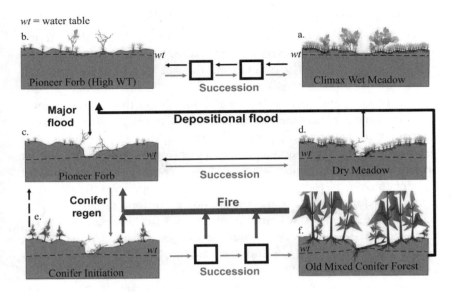

FIGURE 6.4. Potential responses of meadow streams in eastern Oregon to wildfire and old-growth ecosystems. (Wondzell et al., 2007)

the development of wet meadows (fig. 6.4). Good fish habitat tends to occur when the channel is in this condition. The water table can lower if the channel incises over time, drying the meadow and leading to a change in vegetation and a decline in habitat suitability for fish. Impacts of land use, such as grazing, may exacerbate downcutting. If the water table is lowered for an extended period, coniferous forests can invade the area that was formerly too wet for many tree species (fig. 6.4).

Development of an old-growth riparian forest reduces light levels and ultimately primary production. The stream's energy base largely shifts to the input of needles, which tend to be less nutritious than deciduous vegetation for bacteria, fungi, and invertebrates. Thus, old-growth forests along these streams may not necessarily be the most productive for fish. Several recent studies (Dunham et al., 2003; Minshall, 2003) of the effects of wildfire on fish, fish habitat, and aquatic invertebrates in interior western streams have found that there may be positive responses to wildfire in many areas.

Large River Systems

The association of fish with old-growth forests may be stronger in larger streams and rivers low in the stream network, such as the Hoh River on the Olympic Peninsula in Washington (Sedell et al., 1984). Rivers in downstream portions of a drainage basin tend to be lower gradient and set in wide valleys. However, unlike meadow streams of the arid interior or steep headwater streams in the Cascade or Coast Ranges, sediment supplies are less variable because of regular inputs from glacially fed headwaters, and there is a very low likelihood that gravel in these streams would ever become depleted. In such settings, old-growth forests contribute to the complexity and diversity of habitat and productivity of fish by providing large amounts of wood that serve as primary fish habitat structures. The wide valley floors of alluvial rivers and the tendency of floodplain rivers to meander create many canopy openings that maintain aquatic production. As a result, the abundance of food organisms is less likely to be reduced by the dense forest canopy often associated with narrowly confined, old-growth forested headwater streams. Unfortunately, old-growth forests along lowland rivers are rare because of past and present human development.

Higher gradient rivers, in contrast, are much less likely to meander, and they usually contain cobbles and boulders that may persist for long periods of time. In these streams, the aquatic food web is strongly limited by the amount of light reaching the stream (Bisson and Bilby, 1998) and high water velocities that limit algal development. Periodic gaps in the riparian canopy of old-growth forests along these streams, rather than disturbance of large areas, may provide increased light, leading to higher aquatic productivity (Sedell and Swanson, 1984) and greater food availability within the stream for fish.

Conclusions and Challenges

The association among fish, fish habitat, and old-growth forests is not as simple and straightforward as has been assumed and appears to vary widely across the Pacific Northwest and likely across the western United States. Case studies of fish production in Pacific Northwest watersheds tend to support the hypothesis that a range of forest ages, including early and mid-seral stages, leads to a combination of habitat and trophic conditions that are favorable for salmon and trout within the larger drainage system. The potential to extrapolate results from the studies described above to large areas is unknown at this time. We believe that the results are applicable to the watersheds in which the studies were conducted, but it is premature to generalize them over broad areas. The properties of streams over time appear to vary widely depending on local conditions such as climate, geology, topography, and location in the drainage network.

Presumed positive relationships between old-growth forests and fish productivity may hold true in large, floodplain-dominated river systems but may not hold in high-relief streams, where coarse sediment inputs are rare and controlled by very large natural disturbances, such as in the central Oregon Coast Range. Streams in old-growth forests in these settings may not be very productive because of the depletion of gravel and heavy shading, important components of fish habitat.

The association between salmonids and old-growth forests may be strongest in streams that have the capacity to recover quickly from disturbances, such as those studied in the western Cascades, which possess abundant large wood to trap coarse sediment, or in large alluvial rivers lower in the drainage network. However, even in these streams, localized disturbance of the riparian forest is a prerequisite for increased aquatic productivity by creating small openings that allow sunlight to penetrate to the stream as well as excavation and recruitment of buried wood when rivers meander. It must also be remembered that the vegetation dynamics of older forests, for example, can be different between uplands and lowlands, with more frequent and more kinds of disturbances occurring in the uplands. In arid regions, old-growth forests may not necessarily be the most productive setting for fish where factors other than large wood may limit habitat quality. The examples provided earlier suggest that gravel and larger sized material may be lacking when there has been a long interval since the last major disturbance.

We believe that it would be timely to reexamine the assumptions about fish, fish habitat, and old-growth forests by addressing research questions involving multiscale perspectives, from individual habitat units to the ocean. Old-growth forests are one part of this huge range of habitat complexity, but only one part, and we may discover that directing riparian restoration projects to the recovery of old growth is not the sole avenue for capturing the complexity of salmon habitat and life cycle needs. We suggest that landscapes containing a mixture of successional stages may be a more appropriate setting for robust salmon populations. This would require conducting studies involving a range of forest successional stages, not just old growth and clearcuts, which were the focus of many early studies.

In this context, and as opportunities arise, it will be important to reassess the effectiveness of habitat reserves, such as key watersheds in the Northwest Forest Plan, that were designed to make both short- and long-term contributions to the recovery of Endangered Species Act–listed and other legally protected fish populations. Many of these habitat reserves were selected because of the presence of old-growth forests. If, as we have suggested, old-growth forests may not be the most productive environments for fish in some landscape settings, then the overall effectiveness of management and recovery plans for at-risk salmonids may not live up to expectations if the goal is simply to increase the amount of late-successional/old-growth forest stands. Of course, there may be other compelling reasons to increase the amount of old growth (such as habitat for endangered wildlife), but policymakers are best served if the tradeoffs are made clear.

The major challenge to management agencies and policymakers will be to decide whether to approach old-growth forests and streams from a static or dynamic landscape perspective. Each will affect social, economic, and ecological objectives in different ways. The static approach to managing stream habitat targets only one stage of the potential range of conditions in aquatic ecosystems and relies on fixed standards that usually involve setting quantitative targets for environmental attributes—for example, pieces of large wood per unit length of stream. Thus, adopting a static landscape approach may result in two management problems: (i) the long-term depletion of old-growth forests and the benefits they provide to fish if old-growth preserves are spatially "hardened" into land use plans without provisions for their replacement in other areas and (ii) the need for expensive, often frequent, bioengineered substitutes for the large wood and other habitat features provided by old-growth forests to streams. These problems are critical, but they are not often recognized in the current debate. Although there may be some sense of accomplishment in the short term in protecting remaining old-growth forest, it may only postpone dealing with the difficult issue of replacement until the range of potential options is much smaller than what we have today.

A dynamic landscape perspective recognizes that periodic disturbances are needed to maintain ecosystem properties through time and facilitates a natural range of conditions throughout the management area. The dynamic view allows for spatial and temporal variation in environmental attributes, thereby considering their historical or natural range of variability rather than relying on rigid standards. For example, a dynamic landscape approach to managing stream habitat in old-growth forests would include allowing some features characteristic of early, mid-, and late-seral forests, because these patchy conditions result naturally from disturbances in unmanaged areas. It would also involve recognizing that wildfires rarely leave wide buffers around streams: They may burn through them. The relative abundance and size of different forest patches, and their persistence over time, will vary widely across a region. In some places within a watershed, including frequently disturbed floodplains, late-seral forest conditions are rarely attained, and recruitment of large wood occurs when this material is fluvially transported from headwaters during high flows. An actively managed landscape can work toward maintaining many of the desired habitat features of streams in old-growth forests, including their inherent variability, with a variety of forest conditions, both within and among watersheds, some dominated by old forest, others by young forest, and yet others by a mix. This can be done only *if* basic watershed processes are maintained and the habitat-forming legacies of natural disturbances are not disrupted.

Old-growth riparian stands may persist for a long time, often centuries, but most are eventually altered by floods, wildfires, windstorms, insects and diseases, or other natural disturbance agents. Although placing a watershed with abundant old growth into a reserve may achieve the short-term objective of maintaining a forest type that is much less abundant now than in the past, the old-growth features that are so highly desired will not persist indefinitely. Without a strategy that allows forests in other watersheds to attain old-growth properties as existing stands are gradually lost to natural disturbances, river basins will continue to experience a chronic, cumulative loss of large wood and other habitat attributes critical to productive conditions. Nevertheless, maintaining old-growth forests in watersheds may be desirable and even necessary in many drainage systems, but this will not be sufficient to ensure the suite of conditions needed for viable fish populations. In other words, old-growth forest stands, by themselves, do not represent the sole forest age or condition needed for the conservation of native salmon and trout in the Pacific Northwest.

LITERATURE CITED

Benda, L. E., D. J. Miller, T. Dunne, G. H. Reeves, and J. K. Agee. 1998. Dynamic landscape systems. In *River ecology and management: Lessons from the Pacific coastal ecoregion*, edited by R. J. Naiman and R. E. Bilby. New York: Springer.

Bisson, P. A., and R. E. Bilby. 1998. Organic matter and trophic dynamics. In *River ecology and management: Lessons from the Pacific coastal ecoregion*, edited by R. J. Naiman and R. E. Bilby. New York: Springer.

Dunham, J. B., M. K. Young, R. E. Gresswell, and B. E. Rieman. 2003. Effects of fire on fish populations: Landscape perspectives on persistence of native and nonnative fish invasions. *Forest Ecology and Management* 178(1–2):183–96.

Franklin, J. F., K. Cromack, Jr., W. Denison, A. McKee, C. Maser, J. Sedell, F. Swanson, and G. Juday. 1981. *Ecological characteristics of old-growth Douglas-fir forests*.

General Technical Report PNW-118. Portland, OR: USDA Forest Service, Pacific Northwest Forest and Range Experiment Station.

Minshall, G. W. 2003. Responses of stream benthic macroinvertebrates to fire. *Forest Ecology and Management* 178(1–2):155–61.

Murphy, M. L., J. F. Thedinga, K. V. Koski, and G. B. Grette. 1984. A stream ecosystem in an old-growth forest in southeast Alaska: Part V: Seasonal changes in habitat utilization by salmonids. In *Fish and wildlife relationships in old-growth forests*, edited by W. R. Meehan, T. R. Merrell, Jr., and T. A. Hanley. Morehead City, NC: American Institute of Fishery Research Biologists.

Reeves, G. H., L. E. Benda, K. M. Burnett, P. A. Bisson, and J. R. Sedell. 1995. A disturbance-based ecosystem approach to maintaining and restoring freshwater habitats of evolutionarily significant units of anadromous salmonids in the Pacific Northwest. In *Evolution in the aquatic ecosystem: Defining unique units in population conservation*, edited by J. Nielsen. Bethesda, MD: American Fisheries Society Symposium.

Rieman, B. E., and J. D. McIntyre. 1995. Occurrence of bull trout in naturally fragmented habitat patches. *Transactions of American Fisheries Society* 124:285–96.

Sedell, J. R., and F. W. Swanson. 1984. Ecological characteristics of streams in old-growth forest of the Pacific Northwest. In *Fish and wildlife relationships in old-growth forests*, edited by W. R. Meehan, T. R. Merrell, Jr., and T. A. Hanley. Morehead City, NC: American Institute of Fishery Research Biologists.

Sedell, J. R., J. E. Yuska, and R. W. Speaker. 1984. Habitats and salmonid distribution in pristine, sediment-rich river valley systems. In *Fish and wildlife relationships in old-growth forests*, edited by W. R. Meehan, T. R. Merrell, Jr., and T. A. Hanley. Morehead City, NC: American Institute of Fishery Research Biologists.

Wondzell, S. M., M. A. Hemstrom, and P. A. Bisson. 2007. Simulating riparian vegetation and aquatic habitat dynamics in response to natural and anthropogenic disturbance regimes in the upper Grande Ronde River, Oregon, USA. *Landscape and Urban Planning* 80:249–67.

Chapter 7

Contribution of Old-Growth Timber to Regional Economies in the Pacific Northwest

RICHARD W. HAYNES

As battles over old-growth harvest in the Pacific Northwest have ebbed, peaked, and flowed over the past five decades, they have often led and sometimes followed the trajectory of the forest products industry and economies in the region. As one of the regional icons, the forest products industry has evolved from processing old-growth timber to processing timber from forests managed for sustainable forest wood production. Its development provided the livelihood for "dozens of small communities" and helped define the sense of place that frequently motivates residents "to struggle with each other for the future of the lands and homes they love" (Robbins, 1997, 2004).

This chapter traces not just the role old-growth timber has played in the forest products industry and the economy of the Pacific Northwest but also the role it might play in the future. It is a story that includes many of the components defining the old-growth controversy that first emerged in the 1950s and revolved around three related issues: (i) the role and amount of federal (mostly old-growth) timber in timber markets; (ii) the obligations federal agencies have to communities near or among federal timberlands; and (iii) the role forests, especially federal forests, play in regional economies (Smith and Gedney, 1965). This was a period when employment declined in the Pacific Northwest forest products industry, even as harvests remained relatively stable (shown in fig. 7.1a in terms of total harvest and employment

FIGURE 7.1. (a) Forest products employment and harvest and (b) jobs per million board feet of harvest in the Pacific Northwest since 1950.

and in fig. 7.1b as jobs per million board feet of harvest). These issues have remained central throughout the controversies of the 1980s and 1990s.

There are three parts to this chapter: (i) the evolution of the industry; (ii) the evolution of the Pacific Northwest economy for the past thirty-five years; and (iii) the relation between federal forest management actions and the conditions in forest-dependent communities. I close with a brief summary of prospects for the next several decades, when the consequences of recent actions will play out.

The Industry

The forest products industry was among the earliest manufacturing industries to evolve in the Pacific Northwest. It was based on processing lumber for a variety of markets from large trees located near tidewater. The industry

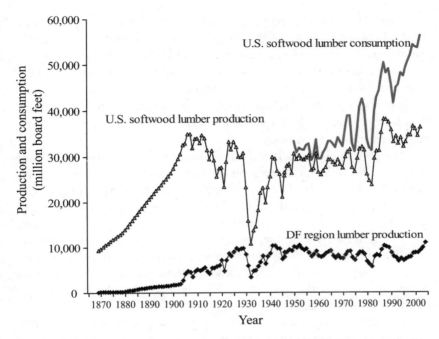

FIGURE 7.2. Lumber production and consumption since 1870 (DF - Douglas-fir).

rapidly expanded after 1900 (fig. 7.2) with the development of railroads. It developed to serve domestic and export markets and was an early adopter of advanced materials-handling and -processing technologies. For the entire twentieth century, mills in the *Douglas-fir region* (defined as the west side of the Pacific Northwest) averaged twenty-eight percent of all lumber produced in the United States, from less than five percent of the timberland base.

Until the late 1940s, production was based on large, relatively clear (knot free) Douglas-fir logs available from private landowners. This lumber commanded premium prices because it was relatively knot-free. Federal logs played a relatively small role because the forest industry attempted to restrict expansions in federal harvests as part of their efforts to improve stumpage prices and forest management on private timberlands (see Mason, 1969, for details).

Federal harvests rapidly increased following World War II as the nationwide housing and industrial expansion pushed softwood lumber demand to new heights. From the late 1940s until the late 1980s, timber harvest in the Douglas-fir region increased roughly twenty-five percent, mostly because of increased harvest on public lands (see fig. 2.1 in chapter 2) (Adams et al., 2006). Between 1945 and 1965, timber harvest on national forest land in the western forests of Oregon and Washington rose from about 745 million board feet (149 million cubic feet) to 4,035 million board feet (807 million cubic feet). During this same period, the proportion of the area of old-growth (more than 160 years old) forests dropped from sixteen to less than four percent on private forest lands and from thirty-six to thirty-two percent on public forest lands.

The growth in domestic and export markets led to rapid rises in stumpage prices (fig. 7.3) and to bans on exporting federal logs without further processing. By the late 1980s, the industry was bifurcated. One rapidly growing segment of highly efficient mills (in terms of volume recovery) cut roughly uniform logs of mostly second-growth private timber for commodity markets. The other segment included several older mills that processed larger (and older) log mixes, mostly from public timberlands, for a range of markets, including high-value domestic and export markets. The design of the older mills made them difficult to adapt to the major changes that would soon shape the industry.

Landmark changes started in 1991 with injunctions on the sale of federal timber that were resolved three years later with the implementation of the Northwest Forest Plan. The resulting reductions in federal sales caused wood supplies to fall below existing processing capacity and led to mill closures, especially those dependent on federal timber, because many could not efficiently process the smaller logs (less than twenty inches) available from private landowners. Demand also fell as the Japanese economy and then other Asian economies collapsed (Daniels, 2005).

Although the changes in markets and federal timber flows required painful adjustments in Pacific Northwest logging and mill communities, U.S. consumers saw little change in product availability or prices as the harvest decline (roughly five billion board feet) was offset by several factors. Some were expected and included increased lumber imports (fig. 7.2); increased private nonindustrial harvests in the Pacific Northwest; and increased harvests in other regions, particularly from managed forests in the South and from mature forests in the interior Canadian provinces (see Haynes, 2003, for a general discussion). Unforeseen was the collapse of the log export market from the Pacific Northwest that allowed timber managers and landowners to shift formerly exported logs (more than 2 billion board feet, log scale, per year) to the domestic market. Also unforeseen was the sustained housing boom from the mid-1990s through 2006.

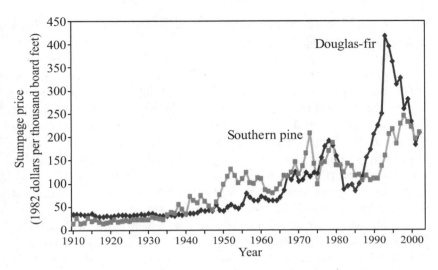

FIGURE 7.3. Stumpage prices for Douglas-fir and southern pine sawtimber since 1910.

The combination of these factors restructured the timber industry in the Douglas-fir region. It is a large industry producing, in 2002, 13.44 billion board feet of lumber in Washington, Oregon, and California and requiring 1.68 billion cubic feet of logs. Much of this production is in highly efficient mills that produce commodity lumber from fourteen- to twenty-inch logs primarily for the domestic market and using timber from private timberlands (see Barbour et al., 2003; Haynes and Fight, 2004). Some of these mills are very large, producing 300,000–400,000 board feet per shift, but there are also specialty mills making products for various niche markets, some of which are relatively small and produce less than 50,000 board feet per shift. There is also an evolving small log industry using logs between 4.5 and 10 inches that may rely in part on fuels management and ecological restoration activities in older forests. Currently few mills are capable of handling logs more than twenty-four inches in diameter, and the large log premium ended in the 1990s. The high product prices accompanying the initial drop in federal harvest stimulated the rapid adoption of substitute products such as glulam beams and wooden trusses for those products formerly made from large logs.

Markets for standing timber (stumpage markets) during the 1990s were highly volatile as landowners and forest products producers adjusted to the reductions in federal timber flows (see Warren, 2004, for various data series and Haynes, 2003, for a discussion of regional and national markets adjustments). Since the mid-1990s, stumpage prices have been declining or stable (fig. 7.3), suggesting reduced financial returns for some forest management practices. These lower prices threaten some landowners' interest in sustainable forest management. Weak stumpage prices, lower future price expectations, and the loss of the export price premium are leading to a shift toward forest management regimes that favor shorter rotations (Haynes, 2005). Under these circumstances, the economic incentive for many private landowners is to grow smaller, more uniform trees, which may widen the gap between ecological conditions on public and private land. These younger forests will not provide the same type of biological diversity as was found in private stands in the past that were managed for larger trees demanded by the export markets.

Finally, the observed employment decline, which started in the 1950s, has turned around, and employment per million board feet is increasing (fig. 7.1b). The original declines resulted as the industry modernized mills and diversified to include high-value log export and plywood industries. During the late 1980s, trends in jobs per million board feet reversed and began increasing as lumber production increased and log exports declined.

Economies of the Douglas-fir Region

Economic conditions have evolved in the Pacific Northwest from the earliest Euro-American settlement in the 1840s, when it was characterized as an Eden (see Robbins, 1997, 2004), to today, when it is characterized by large urban centers with a Starbucks in every mall. This increase in prosperity and in economic infrastructure is reflected in the economic activity in the three related functioning economies (Seattle, Washington, and Portland and Eugene, Oregon) that define the Douglas-fir region (Haynes et al., 1998). Data for the past three decades show that the Douglas-fir region mirrors the economic structure of the U.S. economy, although agriculture plays a slightly larger role in the Douglas-fir region than in the nation as a whole. However, the forest products industry remains about three times more important in the Douglas-fir region than in the United States as a whole, even though in 2003 it accounted for only eleven percent of manufacturing employment, down from nearly eighteen percent in the late 1980s. Northwest job growth is faster than for the United States as a whole, and during the past decade nearly 650,000 new jobs were added there.

The economic structure of both the Douglas-fir region and the United States has broadened over the past three decades with declines in manufacturing being offset by increases in the service sector. This decoupling of the economy from the manufacturing sector is cited as evidence for a new economic paradigm in which economic activity is based on knowledge rather than manufacturing industries (Galston and Baehler, 1995). In the Pacific Northwest, however, the loss of forest products manufacturing jobs associated with reductions of federal harvests has had disproportionate impact on rural communities, especially those adjoining federal lands (see Charnley et al., 2006, for details). At the same time, some rural communities offer lifestyles that attract individuals who do not need employment or have knowledge-based jobs that require little formal infrastructure. Such migration results in economic growth and employment opportunities that may replace lost manufacturing jobs (Power, 1996; Niemi et al., 1999). The Economic Adjustment Initiative that was part of the Northwest Forest Plan also benefited individuals and communities during the period of economic transitions (Tuchmann et al., 1996).

Community Stability

Until the early 1990s, the term *stability* had been associated with jobs and income generated directly or indirectly through the harvest and processing of forest products by residents of communities (SAF, 1989). Many people debated public forest policy assuming that reductions in federal timber flows would reduce local employment opportunities, negatively affect socioeconomic well-being, and threaten the existence of communities. During the 1980s, these debates broadened to include discussions of how communities change and the specific nature of the "social contract" between land management agencies and communities. Some advocates argued that the repeated commitments to local communities embodied in forest-level plans developed in the 1980s and long-standing policies recognized the rights of those who depend on federal forest land for their livelihood.

Research from the past two decades suggests new ways to consider the connections communities have with forest resources. Communities are seen as being dynamic, responding to external factors that induce change. Rather than just considering factors related to employment, other influences, such as connectivity to broad regional economies, community cohesiveness and place attachment, and civic leadership, are now recognized as determinants of community viability and adaptability (Donoghue et al., 2006).

The political compromise leading to the Northwest Forest Plan linked timber production on federal lands with socioeconomic well-being defined at the community level. But the outcome has been mixed: Some communities adjacent to federal forest land have experienced decreases in socioeconomic well-being, but others have found ways to adapt to declines in timber production and other changes in social and economic conditions (see Charnley et al., 2006, for details). At the regional scale, some of the potentially negative economic changes associated with the Northwest Forest Plan were obscured by rapid growth in population and total jobs.

Amid all the larger changes, questions about the nature of the relations among communities, economies, and natural resource management are still prominent in federal land management planning. For example, figure 7.4 shows the 433 communities in the six Bureau of Land Management districts in western Oregon (Donoghue et al., 2006). It suggests that there are more communities in close proximity to Bureau of Land Management and national forest lands than generally perceived. This adds complexity to public agency management goals that seek to support community development or status. It also raises questions about which communities are likely to be affected and about the capacities of communities to deal with largely external changes.

Prospects for the Next Generation

Recent assessments of the U.S. timber situation show a future in which almost all softwood timber after 2015 will be harvested from managed stands on private timberlands, mostly in the South and in the Douglas-fir region (see the discussion [pp. 121–123] in Haynes, 2003). In the past decade, this ongoing transition from harvesting in natural stands to harvesting in managed stands has mitigated some of the consequences of harvest reductions on public lands. In all likelihood the role federal timber plays in the U.S. forest situation will decrease further, enabling federal lands to provide other key goods and services as part of a vast forested commons including both public and private timberlands. There will still be pressures from leaders in communities in close proximity to federal lands who seek greater access to federal resources and more local employment to support economic and community development.

The unique emphasis on protecting older federal forests under the Northwest Forest Plan has a number of socioeconomic consequences. On the positive side the focus on old-growth forests has protected unique (and highly valued) habitat and places both now and for future generations. It has also led to more interest in considering how the

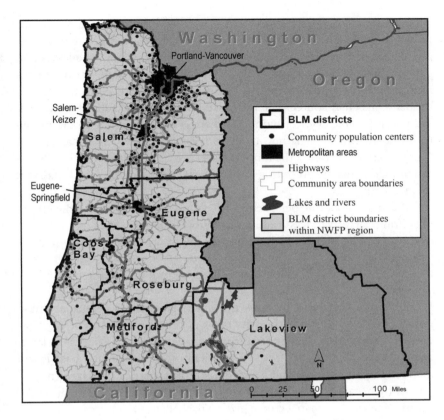

FIGURE 7.4. Map showing communities in Northwest Forest Plan area.

structurally diverse Pacific Northwest timberland base is capable of producing a wide array of environmental services and goods. The focus on old-growth forests has raised questions about how to develop effective ways to govern and manage the set of environmental services and goods across ownership boundaries. It protects many of the uses and values important to urban people, although it does not protect some of the uses and values important to rural people. The increasing emphasis on mature forests could produce both higher quality timber with old-growth kinds of characteristics (large diameter, relatively knot-free, and a high number of rings per inch) and a broader array of environmental services, goods, and conditions, especially in terms of wildlife habitat.

On the unintended, potentially negative side, there are still concerns about how to weigh the ambiguous gains in environmental conditions for almost certain economic losses. There are also ongoing debates around how to protect common property goods and services from abuse (see Hardin, 1968, for the classic discussion). The changes in both the levels and dependability of federal timber flows have increased the investment risks facing the forest products industry, which has restructured to use private timber as its chief source of supply (see Haynes et al., 2003, for details). Public land managers now wonder if there is sufficient local timber industry infrastructure or whether it can be reestablished where it has closed. One current challenge is the limited markets for the small diameter ("nonsawtimber") material resulting from forest restoration and fuel management efforts.

From an economic perspective, the failure to fully examine the effects of shifting harvests to less-well-regulated forests and to other regions demonstrates a policy failure. Although some might say these were unintended consequences, the reality is that the strident nature of the old-growth controversy silenced those asking that policies consider the full range of possible effects.

Difficult to assess are the consequences of landscape-level management goals, given how the wide diversity of ownerships, public and private, has created a patchwork mosaic of management regimes spread across the landscape. In the Douglas-fir region, about half of the timberland base is less than forty years old, and half is more than forty years old, with thirty percent of the total older than eighty years (Zhou et al., 2005). The disparate nature of individual management goals will challenge managers to consider the complementary nature of resource conditions and the contributions of various landowners across the ownership mosaic.

For the past three decades, the growth and evolution of regional economies has altered the economic contributions of old-growth forests. From being the base of timber manufacturing in the late 1970s, old growth by the late 1990s had become an icon helping define what the Pacific Northwest is and offers to residents and visitors. Shifts in societal values have also broadened forest management to consider complex tradeoffs (or compatibility) among a broad set of environmental values, including timber, wildlife habitat, aesthetics, biological diversity, water flows, ecological integrity, and recreation. Finally, the United States is now a nation of urban dwellers who are willing to import forest products from the global economy and who see federal lands as a source of open space and recreation opportunities. It is that role that largely defines the contributions that federal lands make to regional economies.

LITERATURE CITED

Adams, D. M., R. W. Haynes, and A. J. Daigneault. 2006. *Estimated timber harvest by U.S. region and ownership, 1950–2002.* General Technical Report PNW-GTR-659. Portland, OR: USDA Forest Service, Pacific Northwest Research Station.

Barbour, J. R., D. D. Marshall, and E. C. Lowell. 2003. Managing for wood quality. In *Compatible forest management,* edited by R. A. Monserud, R. W. Haynes, and A. C. Johnson. Dordrecht, The Netherlands: Kluwer Academic Publishers.

Charnley, S., E. M. Donoghue, C. Stuart, C. Dillingham, L. P. Buttolph, W. Kay, R. J. McLain, C. Moseley, R. H. Phillips, and L. Tobe. 2006. Socioeconomic monitoring results. Volume III: Rural communities and economies. In *Northwest Forest Plan—The first 10 years: Socioeconomic monitoring results (1994–2003),* technical editor S. Charnley. General Technical Report PNW-GTR-649. Portland, OR: USDA Forest Service, Pacific Northwest Research Station.

Daniels, J. M. 2005. *The rise and fall of the Pacific Northwest export market.* General Technical Report PNW-GTR-624. Portland, OR: USDA Forest Service, Pacific Northwest Research Station.

Donoghue, E. M., N. L. Sutton, and R. W. Haynes. 2006. *Considering communities in forest management planning in western Oregon.* General Technical Report PNW-GTR-693. Portland, OR: USDA Forest Service, Pacific Northwest Research Station.

Galston, W. A., and K. J. Baehler. 1995. *Rural development in the United States: Connecting theory, practice, and possibilities.* Washington, D.C.: Island Press.

Hardin, G. 1968. The tragedy of the commons. *Science* 162:1243–48.

Haynes, R. W., tech. coord. 2003. *An analysis of the timber situation in the United States: 1952–2050—A technical document supporting the 2000 USDA Forest Service RPA assessment.* General Technical Report PNW-GTR-560. Portland, OR: USDA Forest Service, Pacific Northwest Research Station.

Haynes, R. 2005. *Economic feasibility of longer management regimes in the Douglas-fir region.* Research Note PNW-RN-547. Portland, OR: USDA Forest Service, Pacific Northwest Research Station.

Haynes, R. W., and R. D. Fight. 2004. *Reconsidering price projections for selected grades of Douglas-fir, coast hem-fir, inland hem-fir, and ponderosa pine lumber.* Research Paper PNW-RP-561. Portland, OR: USDA Forest Service, Pacific Northwest Research Station.

Haynes, R. W., W. J. McGinnis, and A. L. Horne. 1998. Where are the jobs? Lessons learned about the economic impacts of ecosystem management. In *Forest policy: Ready for renaissance,* edited by J. Calhoun. Seattle: College of Forest Resources, University of Washington.

Haynes, R. W., D. M. Adams, and J. R. Mills. 2003. Contemporary management regimes in the Pacific Northwest: Balancing biophysical and economic
concerns. In *Compatible forest management*, edited by R. A. Monserud, R. W.
Haynes, and A. C. Johnson. Dordrecht, The Netherlands: Kluwer Academic
Publishers.

Mason, D. T. 1969. Memoirs of a forester: Part III. *Forest History* 13(1&2):28–
39.

Niemi, E., E. W. Whitelaw, and A. Johnston. 1999. *The sky did NOT fall: The
Pacific Northwest's response to logging reductions.* Prepared for the Earthlife
Canada Foundation and the Sierra Club of British Columbia. Portland, OR:
ECONorthwest.

Power, T. 1996. *Lost landscapes and failed economies: The search for a value of place.*
Washington, D.C.: Island Press.

Robbins, W. G. 1997. *Landscapes of promise: The Oregon story, 1800–1940.* Seattle:
University of Washington Press.

Robbins, W. G. 2004. *Landscapes of conflict: The Oregon story, 1940–2000.* Seattle:
University of Washington Press.

SAF (Society of American Foresters). 1989. *Report of the Society of American Foresters
national task force on community stability.* SAF Resource Policy Series 89–06.
Bethesda, MD.

Smith, R. C., and D. R. Gedney. 1965. *Manpower use in the wood products indus-
tries of Oregon and Washington 1950–1963.* Research Paper PNW-28. Portland,
OR: USDA Forest Service, Pacific Northwest Forest and Range Experiment
Station.

Tuchmann, E. T., K. P. Connaughton, L. E. Freedman, and C. B. Moriwaki. 1996.
The Northwest Forest Plan: A report to the president and Congress. Portland, OR:
USDA Forest Service, Pacific Northwest Research Station.

Warren, D. D. 2004. *Production, prices, employment, and trade in Northwest forest
industries, all quarters 2002.* Resource Bulletin PNW-RB-241. Portland, OR:
USDA Forest Service, Pacific Northwest Research Station.

Zhou, X., R. W. Haynes, and R. J. Barbour. 2005. *Projections of timber harvest in
western Oregon and Washington by county, owner, forest type, and age class.* General Technical Report PNW-GTR-633. Portland, OR: USDA Forest Service,
Pacific Northwest Research Station.

Chapter 8

Sacred Trees

ROBERT G. LEE

The birth of "old growth" as the iconic forest can be encapsulated in a few words describing social meanings, time, and space: re-enchantment trumped rationality; the eternal present absorbed the chronology of forest growth; mystical places colonized the choreography of sustained yield operations. The ruling worldview of forests changed dramatically. Although coming to many as an unwelcome shock and to others as a revelation, new meanings of old trees soon took hold of popular culture. A cultural event of such magnitude is infrequent and unanticipated and, as such, invites further examination. What happened? Why did it happen? What are the cultural implications? In this chapter, these questions will be answered from a sociological perspective, particularly considering changes in secular and religious attributes of time and space.

Forests in a Postmodern World

Rationality and science defined the ruling view of forests throughout most of the twentieth century. Beginning with the progressive movement (late 1800s to early 1990s) (Hays, 1969), professional foresters were trained in science and rational decision making and given responsibility for first discerning the public interest in forests and then managing forests to secure these interests. Values were assumed to be separable from facts. Foresters adopted the economic logic of rational comprehensive planning and developed procedures for rationally selecting means to achieve desired ends. Forestry schools trained professionals to use rational methods for bringing order not just to forests but, to the extent possible, to the local communities depending on forests for their livelihood. Foresters were given a "toolkit" equipping them to act as agents for the state or corporation. After World War II, this toolkit featured economic optimization models for scheduling forest harvests and simulation models for conserving endangered species. The rational approach reached a head with the regulations for implementing the National Forest Management Act of 1976. It required that multiple forest values be reduced to a dollar value and decisions be made by a linear programming model (FORPLAN).

Foresters trained as rational agents for the state or corporation were poorly prepared for social changes arising toward the end of the twentieth century. They were confused when emotional reactions revealed that stakeholders could not separate facts from values. They retreated in anger when their professional authority was questioned by people with diverse and conflicting claims on new and emerging forest values. They rejected the claim that rational thinking was based on the illusion—the faith—that rationality could bring order to nature as well as to human society. In short, they failed to comprehend the emergence of what some sociologists call "postmodern society."

Zygmunt Bauman (1993: 32), a British sociologist, talked about postmodern society as

> modernity without illusions. The illusions in question boil down to the belief that the 'messiness' of the human world is but a temporary and repairable state, sooner or later to be replaced by the orderly and systematic rule of reason.

He said that the modern world we knew for most of the twentieth century failed to accept the truth that the messiness of the world is part of the human condition. Foresters trained in the twentieth century epitomized this faith in reason and felt their world had been turned upside down. They were committed to bringing order to the forest and replacing the messiness of "decadent" older forests with manageable, fast-growing plantations of uniform trees.

Foresters were most troubled by what can best be described as the re-enchantment of the natural world. Bauman (1993: 33) described how modernity unsuccessfully sought to disenchant the world when he said,

The mistrust of human spontaneity, of drives, impulses, and inclinations resistant to prediction and rational justification, has all but replaced the mistrust of unemotional, calculating reason. Dignity has returned to emotions; legitimacy to the inexplicable, nay irrational, sympathies and loyalties that cannot explain themselves in terms of their usefulness and purpose. . . . The postmodern world is one in which mystery is no more a barely tolerated alien awaiting a deportation order.

Re-enchantment of nature found expression in a sudden shift in the cultural meanings associated with old trees. Old forests became "old growth" or "ancient forests." *Old growth* now connotes a world of mystery and stirs deep emotions. As such, it does not make sense to those whose thinking is ruled by reason. An understanding of how people experience space and time will help answer the question of how the meaning of trees took an unexpected turn toward mystery and emotion.

Foresters' Time and Space

The model forest—referred to as the normal forest—was to foresters what perfect markets are to economists, Bach is to musical composition, or national champions are to dog show people. It was a standard against which to measure the common realities of forest structure and composition. Few forests approximated this standard, but all were shaped by aspirations to attain it. But today the normal forest no longer serves as the common standard for measuring success in forest management because "old growth" was successfully asserted as a rival ideal. What was the so-called normal forest, and which of its properties made it vulnerable to challenge?

The goal in managing the *normal* forest was to harvest wood so that an annual volume of cut would be replaced by annual growth. By the nineteenth century, German and French foresters had developed methods to calculate rates of growth and prescribe harvest levels. Here is a simple example. Presume it takes 100 years of growth for trees to reach harvestable age. A forester would divide the forest into 100 compartments and cut the existing forest and plant new trees on one compartment each year. After 100 years, the forest would be perfectly regulated, with one compartment reaching maturity that year. The next year, the next oldest compartment would reach 100 years of age and be ready for harvest and regeneration. This practice was thought to ensure "sustained yield" production of wood products for perpetuity.

The idea of the normal forest epitomizes the rational thinking of the modernist era. The messiness of natural forests was to be ordered by forest regulation. Manipulation of both time and space was fundamental to bringing order to forests. Considerations of time involved measuring how much trees of a given species grow each year and calculating volume accumulation to predict harvestable age. Space was derived from time by calculating how much area of a forest should be harvested and regenerated each year to set a harvest level that would ensure a constant supply of wood.

The modernist's conception of time was chronological time—the time of duration. As such, foresters' time was measured in the annual biological cycle of years. The postmodernist's conceptions of time, as described below, are not restricted to chronological time, allowing expressions of the timelessness of special places and cyclical time of seasons and biological processes.

Modernist conceptions of space emphasized utilitarian values. Like years on a chronological sequence, spaces were interchangeable. One place had significance only because of how it could be compared with the utilitarian values of another place. Homogenized space was reflected in the hyperrationality of the FORPLAN model mentioned above. Modernist national forest planning treated all available acres as having the same potential value and used a computer to assign them economic values. Postmodernist conceptions of time are more encompassing than utilitarian values and can reflect the unique or special value of places that stand beyond chronological time. The concept of inherent value, and even spiritual value, anathema to the modern rationalist, has been openly invited to enter the postmodernist pantheon.

"Old-Growth" Space and Time

"Old growth" may be as much a manifestation of spiritual yearnings as a mandate for protecting biodiversity. Rational strategies for protecting endangered species are intertwined with symbolism that can only be described as religious. But why do old trees or undisturbed forests attract such rich meanings? Part of the answer may be found by contrasting old growth with not only the forester's rationalized world but also the modern world from which the managed forest emerged.

The triumph of neoclassical economic rationality epitomized modern society. Time became money. Cycles of seasons, weeks, and days, marked by festive celebrations, were replaced by homogeneous time in which moments were reducible to the same monetary values. The Sabbath was at first redefined as a day free from the demands of work and then soon became a workday when it became a time for commerce. Private enterprise colonized the night with twenty-four-hour supermarkets. Work brought money, and money fed ever-increasing consumption of goods and services.

Space also became money. The special significance of places was replaced by near-universal recognition of their monetary value. Native claims that particular places had special powers were ignored by land managers. Homes and farms became investments and were valued accordingly. Forested lands were valued for their commercial potential. But not all places were reduced to monetary value. National parks, state and local parks, historic sites, and wilderness areas were reserved to mark them off from commercialized space.

"Old growth" is perhaps best defined as a symbolic refuge from an increasingly commercialized world. The market mentality dominated policymaking in the United States throughout the modernist era and remains a dominant justification for policymaking. Corporate capitalism has become so intertwined with government decision making that there is little distinction between a public realm and a private realm, and serving business interests is generally seen as the best means for advancing public interests. Social services as diverse as prisons, education, and medical care adopted the market as a model for making decisions. Traditional sources of truth take second place behind worldly concerns with political power or wealth accumulation. Churches model themselves after businesses and increasingly emphasize market share and use business techniques to attract and maintain a tithing membership. The postmodernist value of old growth as a symbolic refuge from commerce was recently rivaled by the language of the market when protection of forest ecosystems was justified on the basis of the economic value of "ecosystem services" performed by forests—a sign that the postmodernist challenge can still be trumped by economic rationality.

Visiting old growth is not required to appreciate its meaning, because this place is first of all a refuge for the imagination, not a material condition. It is a place of power because natural processes are free to function unimpeded by human demands. As such, it opens a door to different experiences of space and time.

Religious imagery abounds in discussions of old growth. Old-growth trees are imagined to be large and tall. Their lifespans exceed that of humans by centuries, so they serve as symbols of immortality. They embody the realms of the earth and spirit, because they are rooted deep in the earth and are so tall they touch the heavens.

The power people find in forests is not new. Van der Leeuw (1963: 394) quoted and commented on Ovid's *Fasti* to reveal experiences of forests that are as alive with meaning today as they were for the Greeks and Romans: "Under the Aventine there lay a grove black with the shade of holm-oaks: at a sight of it you could say there is a *numen* here."

Even today we remain conscious that the gloomy forest has numinous character, although to a great extent we have destroyed primitive man's dread of the seat of the powers by our romantic twilight moods. Nevertheless, the sacred grove arouses its shiver of fear as well of ecstasy.

> If ever you have come upon a grove (*lucus*) that is full of ancient trees, which have grown to an unusual height, shutting out a view of the sky by a veil of pleached and intertwining branches, then the loftiness of the forest, the seclusion of the spot, and your marvel at the thick unbroken shade in the midst of the open spaces, will prove to you the presence of deity (*fidem tibi numinis faciet*) (Van der Leeuw, 1963: 401).

Old-growth forests, unlike houses, temples, or settlements, are not selected and ritually blessed by civil or religious ceremonies. They are instead discovered or revealed. Their position is provided by nature, and they acquire the meanings of old growth when people find a position situating them outside the everyday world. Not all old trees constitute old growth: Groups of old trees in parks and patches in fragmented landscapes may not reorient the finder to the world. But like the sacred trees and groves in other cultures, "old growth" elicits awe and a deep sense of humility in the face of nature.

Although "old growth" is rooted in imagination, and need not be visited to provide meaningful experience, its symbolic power can be renewed by periodic pilgrimages. Van der Leeuw's (1963: 401) comments about pilgrimages seem to apply as well to visiting old-growth forests:

> Primitive man's longing for house and home, for his native country or town, is ultimately therefore the yearning for salvation, for the consciousness of powerfulness, bestowed by one's own selected place . . . man feels happy only in the place which he has discovered, which has stood the test for him and with whose power his own power is associated.

Old-growth forests are also free from demands of commoditized clock time, with its treatment of all time as homogeneous and reducible to money. Past and future collapse into a moment in which meaning is concentrated and distilled into an experience of timeless wonder and awe.

Such an encounter with the sacred is foreign to everyday experience in a commoditized world.

Unlike commoditized time, typified by rates of return on investment, "old growth" assumes the mantle of reversible and cyclical sacred time (Eliade, 1959). "Old growth" is recoverable and indefinitely repeatable. The reversibililty of "old-growth" time provides the cultural template and social legitimization of efforts at forest restoration. From a postmodernist perspective, forest restoration can be viewed as a symbolic event to *restore* the beginnings—often imagined in the United States as the landscapes that prevailed at the time of European conquest. In this sense, forest restoration is about myth-making in a world in which the rich meanings provided by myths have been supplanted by the one-dimensional world of money-making. Only within the illusionary world of late-modernism could forest restoration be justified as a rational, scientific enterprise.

The temporal rhythms of old-growth forests provide opportunities for encountering sacred time—particularly times of renewal and salvation. Fire, wind, flood, and volcanic disturbances are often symbolized as times of rebirth. A highly successful National Park Service public relations campaign propagated rebirth as a primary meaning following the Yellowstone fires of 1988. Instead of destroying the scenic beauty of the park, these disturbances are seen as giving new life to the forest. Rather than being seen as death and destruction, these events are enlivened with the saving power of renewal—of living again. The mythical Egyptian symbol of death and renewal, the phoenix, is often called upon to make sense out of large forest disturbances, particularly fires. The dynamics of old-growth forests are richly decorated with meanings of salvation.

Making meaning out of natural disturbances often involves conflating seasonal cycles and natural biological cycles. Recent conflict over postfire recovery of forests burned in the Biscuit Fire in southern Oregon in 2002 provided new opportunities for likening forest recovery to the coming of spring. The symbolic meaning of spring draws on deep roots extending far back in history to when agricultural people experienced the year as the year of salvation, with the coming of spring and summer bringing fruit that was "indissolubly linked with the saviour" (Van der Leeuw, 1963). Premodern people ensured salvation by adhering to rituals and celebrations that honored the power of nature and its spirits. In addition to the hardy few for whom hiking deep into old growth is a renewing pilgrimage, postmodern society seeks salvation in old growth with rallies, demonstrations, coffee-table picture books, congressional hearings, and television and movie images.

Cultural Implications

We live in a time of uncertainty in which the prevailing social myths are collapsing. Enlightenment faith, with its emphasis on rationality and science, no longer integrates society. People have experienced a weakening of the central principles that guided their lives, resulting in confusion and disorientation. The sociologist James Aho (1997) referred to this crisis as the "apocalypse of modernity." People found a firm sense of identity and confidence in the Enlightenment thinking and white, male Eurocentric culture that characterized the modern era. The modern era is coming to an end, and there is no unifying vision or set of myths to replace Enlightenment thinking. What sociologists call postmodern society may simply be a manifestation of a transition to a new social and economic order. The resulting search for unifying meaning and purpose leads many people to seek a return to origins. Others find meaning in diversity and openness to change.

Aho argued that few people are capable of embracing the postmodern diversity and uncertainty and that fundamentalism is an attempt to reconstitute original meanings. He noted the rise of religious fundamentalism throughout most of the world. Armstrong (2000) was even more thorough in emphasizing religious fundamentalism as a quest for original meaning and included Islamic, Jewish, and Christian fundamentalism in her analysis.

None of these scholars noted the parallels in thinking found among those who seek meaning in reconstituting the original meaning of nature. Given its emphasis on what has been described as sacred space and time, old growth may be an expression of this longing for home and place, if not a longing to stand in relationship to a sacred "other" (Cronon, 1995). Yet old growth, by combining sacred groves with ecological rationality, is unlike other attempts to return to original states of being. Old growth is a refuge for both the human spirit and a diversity of species threatened by the advance of commercialized landscapes. Will its marriage with scientific rationality ensure that old growth remains a place reserved for celebration of both biological diversity and the power of nature? Will this marriage provide a position from which to survey the advancing forces of commercialism and the struggle for limits to a market-based society?

The postmodernist perspective suggests that society is currently encountering old growth as both subject and object—as both a sacred "other" and a biological reality. As Aho suggested, this is particularly messy and creates tension both within and between individuals. Sociologists are not of one mind about the emergence of postmodernism. If it is a transitional state, then the marriage of scientific rationality and sacred groves may dissolve and give way to some unforeseen world of the future ruled by a resurgence of Enlightenment thought or some brand of fundamentalism. If not transitional, then this marriage will come to reflect a society in which people learn to tolerate the ambiguities and messiness of life by developing the capacity to experience the power of old growth without sacrificing the rigor of the scientific method. But what we do know is that the future of old growth, as with all of nature, is ultimately contingent on emergent and ever-changing social worlds.

LITERATURE CITED

Aho, J. A. 1997. The apocalypse of modernity. In *Millennium, messiahs, and mayhem: Contemporary apocalyptic movements,* edited by T. Robbins and S. J. Palmer. New York: Routledge.

Armstrong, K. 2000. *The battle for God.* New York: Alfred A. Knopf.

Bauman, Z. 1993. *Postmodern ethics.* Malden, MA: Blackwell Publishers, Ltd.

Cronon, W. 1995. The trouble with wilderness, or getting back to the wrong nature. In *Uncommon ground: Toward reinventing nature,* edited by W. Cronon. New York: W.W. Norton and Company.

Eliade, M. 1959. *The sacred and the profane: The nature of religion.* New York: Harcourt, Brace, and World.

Hays, S. P. 1969. *Conservation and the gospel of efficiency: The progressive conservation movement, 1890–1920.* New York: Atheneum.

Van der Leeuw, G. 1963. *Religion in essence and manifestation,* vol. 2. New York: Harper & Row.

Chapter 9

Old Growth and a New Nature: Ambivalence of Science and Religion

JIM PROCTOR

This is a chapter about ideas, specifically the ideas we invoke to justify—to ourselves and/or others—saving nature. An emphasis on ideas may seem incongruous in a book about old-growth forests, but what we will be doing is not so much thinking about old growth as thinking about *thinking about* old growth. Ideas are like vehicles: Some get you farther than others, some allow navigation through varied terrain, and some are much more beautiful than others. I suggest that the better we think about thinking about old growth, the farther we'll go in saving nature, the more varied terrains of nature we will successfully navigate, and the more beautiful our journey will be.

A key to this journey comes in realizing that the ideas we invoke to justify our treatment of nature often involve a mixture of science and religion. Consider environmentalist appeals on the public stage regarding old-growth forests of the Pacific Northwest. This outreach often invokes science to demonstrate that old-growth forests are ecological gems, eminently worthy of protection. But there is something more, something deeply felt in the beautiful imagery accompanying many old-growth publications, something violated by depictions of clearcut forest landscapes. Perhaps spiritual sentiment is not explicitly brought forward in public environmental discourse relative to scientific facts, but nonetheless you comprehend that these forests are not just materially significant as sources of carbon sequestration, hydrological regeneration, faunal habitat, and so forth. They are, more fundamentally, *spiritually* significant; above all others they can claim the title of sacred groves, and whether you consider yourself religious or not, their sacredness is—at least as much as the apparent facts of environmentalist outreach—indisputable.

The combined effect of these two dimensions is far more powerful than either component alone. Environmentalists could defend old-growth forests as critical stores of ecosystem services, but this is not an argument that moves people. Environmentalists could, conceivably, defend old-growth protection using only the sacred groves argument, but they would quickly be dismissed as preachy and impractical. Albert Einstein once famously observed (1954), "Science without religion is lame, religion without science is blind"; I would submit that, in the old-growth case, science without religion is sterile, and religion without science is quixotic.

From the use-whatever-works viewpoint, the more the merrier: If we can best defend old-growth protection by appealing to science *and* religion, then by all means let us do so. One immediate objection to this approach is that science and religion can readily be invoked to support rather different policy ends. After all, forest management regimes responsible for old-growth decline have generally championed their reliance on scientific rationality, and the religious zeal of manifest destiny was long a resonant tone in the subduing of American forests (Williams, 1989).

Citing this back-and-forth of invoking science and religion to defend whichever policy one supports is usually followed by some sort of critical comparison to determine who has the high ground. But this is not what I intend to do here. I wish to demonstrate that both have played key roles in our understanding of environmental issues such as the old-growth controversy in the twentieth century. I will do so by means of empirical data I gathered in 2002 in a nationwide survey. My findings suggest that science and religion are both deeply tied to American environmentalism in general.

These findings challenge us to reevaluate the trilogy of science, religion, and environmentalism, perhaps nowhere more so than with that quintessential icon of late twentieth-century American environmentalism, the old-growth forest. As we struggle to understand and evaluate our environmental impacts from the past century and come to terms with how we should collectively act in the next, we need equally to rethink the roles of science and religion in informing our relations with nonhuman nature. In so doing, we may discover a new iconography of old-growth forests, one respectful of their potent symbolism as both sacred groves and ecological treasures. However, we need also to be more willing to acknowledge the long shadows of these scientific and religious icons and the complexities of blending them so freely.

Science, Sacred Nature, and American Environmentalism

The standard story of recent American environmentalism is that it is an outgrowth of scientific discoveries of a natural world imperiled by human practices: Think, for instance, of Rachel Carson's *Silent Spring* (1962), a vivid account of the deleterious impacts of the pesticide DDT. There is no doubt that scientific documentation of human impacts has played a critical role in contemporary environmentalism. But what of religion, of our sense of the sacred and its effect on how we live our lives? A 2002 nationwide survey I conducted suggested, surprisingly, that religion is as much a feature of contemporary environmentalism as science.

The survey involved a sample of roughly 1,000 American adults, screened to represent the U.S. population by sex, age, and region. We included questions that gauged respondents' level of environmental concern in three ways: (i) To what extent did they self-identify as an environmentalist? (ii) How concerned were they about major environmental issues such as air quality or biodiversity loss? (iii) How much did they engage in typical proenvironmental behaviors such as practicing energy conservation or giving money to related causes? Other questions had them appraise their background in and affinity with science: How much science did they study in school? What do they think of scientific rationality?

A more complex set of questions was asked about religion because religion is a concept with many meanings. In addition to questions examining religious beliefs and behavior, what I was especially interested in was the notion of sacredness in nature. Preliminary analysis suggested that environmentalists maintained one of three general notions: (i) the secular idea that nature is important but not sacred; (ii) the broadly Judeo-Christian idea that nature is sacred as created by God; and (iii) the idea that nature is inherently sacred, one generally understood as inconsistent with our religious and secular Western traditions. Interestingly, this last approach seemed to be especially prevalent among environmentalists, but no rigorous study had yet been done to establish its significance.

Based on interviews, I came up with fifteen candidate statements representing a spectrum of opinion on sacredness in nature and in a preliminary survey narrowed them down to six; samples included "Nature is the handiwork of God," "Nature has an important spiritual dimension to it," and "Nature is the result of material forces, not God." These six were empirically boiled down to two primary variables, using a technique called *factor analysis*. The first put the secular and the Judeo-Christian notions on opposite ends of a spectrum. The second concerned inherent sacredness in nature.

Data from the survey allowed me to gain some statistically representative information about American environmentalists. Let's focus on those who self-identified as environmentalist: This approach has limitations (e.g., we do not know whether a person who identifies as environmentalist actually "walks the talk"), but to the extent that the shoe fits we can expect some measure of sympathy and support for environmentalist values and practices. The top portion of table 9.1 suggests a surprising result: Although we usually think that environmentalists come from a distinct sector of society, simple correlation analysis reveals very weak associations between environmentalist self-identification and demographic characteristics, including age, gender, education, and income. This means that Americans who think of

TABLE 9.1. Correlations between identification as an environmentalist and demographic, science, and religion characteristics.

Characteristic	Correlation[a]
Demographic	
Age	0.091
Education	0.081
Income	0.063
Gender	—
Science	
Self-identification as rational	0.089
Training in science	0.270
Trust in science	0.288
Religion	
Self-identification as religious	−0.068
Self-identification as spiritual	0.181
Belief in God	−0.121
Trust in religion	−0.074

NOTE: U.S. adult survey, June–July 2002 ($N = 1,013$).
[a] All listed correlations significant at $p \leq 0.05$.

themselves as environmentalist are not necessarily young or old, male or female, well or poorly educated, rich or poor.

Although American environmentalism apparently cannot be explained by demographic characteristics alone, to what extent can it be viewed as an outgrowth of scientific rationality and/or spiritual impulse? Table 9.1 presents selected results of correlations between environmentalist self-identification and measures related to science and religion. As regards science, environmentalists don't particularly think of themselves as more or less rational, but there is a moderate positive correlation with background in science as well as trust in scientific knowledge for guidance in life, so we do see some proscientific characteristics of environmentalists. These correlations, although not strong, are nonetheless stronger than those related to religion, which are quite varied: Environmentalists don't necessarily self-identify as religious, only weakly identify as spiritual, tend only weakly to believe less in the existence of God, and place no more or less trust than others in religion for guidance in life. These findings are surprising as well: They tell us that there are many forms of religious preference among environmentalists in the United States.

So far, American environmentalism seems more an outgrowth of science than of religion. But this is not the full story: Comparing tables 9.1 and 9.2, the highest correlation with environmentalism concerned neither science nor standard measures of religiosity but rather the respondent's attitude toward inherent sacredness in nature (a factor built from their response to several related questions). Table 9.2 correlates support for inherent sacredness with all three measures of environmentalism noted above, suggesting strong associations with each.

Table 9.2 includes both uncontrolled correlations and partial correlations designed to minimize the possible effect of other factors. Considering these other factors is crucial to avoid jumping to spurious conclusions: What if environmentalism is prompted not so much by a belief in sacredness of nature per se but perhaps by some other factor (e.g., level of education or political orientation) that itself is associated with higher levels of belief in sacredness of nature? This possibility was tested via linear regression analysis (table 9.3), in which major candidate factors generally thought to affect environmental concern—demographic characteristics, political orientation, and theological conservatism—were entered into the analysis first, and only then was the attitude of nature as inherently sacred added to determine its explanatory power with all these other factors taken into consideration.

TABLE 9.2. Correlations between belief in sacredness of nature and environmental characteristics.

Correlation type	Environmentalist self-identification	Environmental issues concern	Proenvironmental behavior
Zero-order[a]	0.303	0.395	0.339
Partial[b]	0.274	0.362	0.303

NOTE: All correlations significant at $p < 0.001$.
   [a]A zero-order correlation ignores the values of other variables.
   [b]Partial correlations controlled for demographics (age, education, gender, income), political orientation, and theological fundamentalism.

The results were surprising: Even following introduction of other possible explanations of environmentalism, the belief among environmentalists that nature is inherently sacred was so strong that it accounted for roughly half of all differences explained by these factors.[1] A comparison of beta values, which indicate how much each factor explains differences in environmentalism, in table 9.3 suggests the strong explanatory power of this belief. What this all means is that no matter one's age, education, gender, or income, no matter one's political or theological orientation, there is a highly powerful factor we can use to predict the level of environmental self-identification, concern, and (reported) practice among Americans, and it is the belief that nature is sacred.

If this belief and related practices constitute a religion of sorts, as one prominent religious scholar has already argued (Albanese 1990, 2002), then American environmentalism most definitely is a religion as well as a science. The historian Lynn White ironically prescribed a religious solution in his famous indictment of religion as the source of environmental problems (White, 1967; Proctor and Berry, 2005); maybe one need look no further than contemporary American environmentalism.

---

1. Total variance explained by all factors was between seventeen and twenty percent—not high, but not an unknown range in social sciences regressions. It should be noted, in comparing the relative contribution of other factors, that political orientation and theological conservatism were highly correlated ($R = 0.343$), so adding political orientation to the regression analysis prior to theological fundamentalism greatly diminished the marginal explanatory power of the latter.

TABLE 9.3. Results of regression of three measures of environmentalist association (self-identification, concern for issues, and behavior) and demographic and political characteristics, and beliefs about the Bible and nature.

	Environmental self-identification			Environmental issues concern			Proenvironmental behavior		
	Beta[a]	R[b]	R^2 Total	Beta	R^2	R^2 Total	Beta	R^2	R^2 Total
1. Demographic characteristics		0.016			0.023			0.020	
Age	0.112**			0.128***			—		
Education	—			—			—		
Gender	—			—			−0.062*		
Income	0.058*			—			—		
2. Political orientation		0.072			0.058			0.073	
Conservative vs. liberal	−0.194***			−0.169***			−0.208***		
3. Theological fundamentalism		0.010			0.002			0.003	
Biblical literalism	−0.097**			—			—		
4. Nature sacredness		0.068			0.120			0.083	
Immanent sacredness	0.267***			0.357***			0.296***		
			0.166			0.197			0.179

NOTE: Independent variables entered as blocks in sequence as above. Political orientation and theological fundamentalism highly correlated ($R = 0.343$), thus order of entry into regression reduces explanatory power of fundamentalism.

[a]Beta represents the amount that the dependent variable (environmentalism measure) changes when the independent variable (e.g., demographic) changes by one unit.

[b]R^2 is the fraction of the variation in the dependent variable that is explained by the independent variable(s).

$^*p < 0.05$; $^{**}p < 0.01$; $^{***}p < 0.001$; results omitted where $p \geq 0.05$.

Old Growth and a New Nature: Embracing Ambivalence

These empirical results offer significant implications for how we character-ize and interpret environmental sympathies and behaviors among Ameri-cans. There can be little argument that both science and religion (the latter founded on a deep belief in nature as inherently sacred) are powerful moti-vating forces in the environmental passions that made old growth such a central American policy issue in the 1980s and 1990s.

Yet what do the results of our survey of American adults mean in terms of policy disputes surrounding old-growth forests? For the avowed old-growth protector, they must sound pretty good: In terms of ecological peril and sacred resonance, many sympathizers would say, you can't get much better than an old-growth forest. Indeed, the surprisingly strong connec-tion between environmentalism and notions of the sacred in nature yields a clear policy imperative of setting aside old-growth forests from logging. The reason is evident in the history of the term *sacred*, of which one core defini-tion found in the *Oxford English Dictionary,* tracing back nearly four centu-ries, is "Dedicated, set apart, exclusively appropriated to some . . . special [or religious] purpose." In the Western sense of the word, then, sacredness trumps utility. You do not ask, "How much old growth do we need?" You ask, in effect, "Is it sacred?" If so, it must be set aside. Such overtly religious language is, of course, rarely used in the policy arena, but whether you look at environmentalist outreach or nationwide survey results, the significance of the sacred is undeniable.

To the environmentalist who is not unduly troubled by the above, let me trouble you a bit. Once we justify protecting old-growth forests for their near-unsurpassable ecological and spiritual value, what other land-scapes would qualify for similar protection? What do we do about all the more ordinary landscapes, those we cannot set aside, those of rather plain ecological qualities? Perhaps you have won the battle in saving old growth but will lose the war on biophysical nature. In a similar vein, Michael Pollan once wrote, "We have divided our country in two, between the kingdom of wilderness, which rules about eight percent of America's land, and the kingdom of the market, which rules the rest" (Pollan, 1991: 189).

As further trouble: Although science and religion seem to mesh conve-niently in the case of old-growth forests, they may not always mesh in the minds of environmentalists. In our study, the correlation between back-ground in science and a sense of nature as inherently sacred was statistically insignificant, implying that environmentalists come in all stripes. There are all-science-no-religion environmentalists, there are all-religion-no-science environmentalists, and there are science-and-religion environmentalists, as well as shades of gray between these extremes.

Each alternative has its downside. Consider, as a starter, the inclusive, integrative view that both science and religion point commonly to a unified vision supporting protection of nature. This view has enjoyed a considerable amount of recent support, evidenced for instance in several large joint statements of environmental concern signed by major world scientists and religious leaders (e.g., Mission to Washington, 1996; National Religious Partnership for the Environment, 2004). Yet the juxtaposition of the languages of science and the sacred leads to some difficult questions. What do you get when you mix the two: Spiritual ecology? Rationalist neo-Romanticism? Alhough the Internet abounds with self-proclaimed visionaries who weave science and religion together into seamless metaphysical and rhetorical wholes, most scientists I know would be less than comfortable with this outcome, and I take their concerns seriously.

A weaker form of support for both science and religion as inspirations for environmentalism, evidenced in the joint statements noted above, views them as separate but equally essential, much as the famous Einstein quote. Here the desire is not to unite science and religion but to create a tidy separation between them so that they may each serve significant but nonoverlapping roles. This view was formalized by the late Stephen Jay Gould, who argued that science and religion constitute "non-overlapping magisteria" of authority over the domains of facts and values, respectively (Gould, 1999). To think, however, that science is all about facts, and religion all about values, is to chop off the feet (perhaps even the heads!) of both on this binary Procrustean bed. To deny that values are embedded in science is to deny the cumulative scholarship of the history, philosophy, and sociology of science. To ignore the factual assertions—whether verified by empirical evidence or faith—of religious movements is to ignore some of their fundamental claims. Facts and values are separate only in a highly abstracted, fairyland version of reality, as I've argued at length elsewhere (Proctor, 1998a, 1999, 2005). No, science and religion bump up against each other precisely because they cannot readily be sent to their purified corners.

There is another alternative: Simply do away with religion (or science), and environmentalism will be better off for having done so. Thus, Paul and Anne Ehrlich (1996) have warned how religion may "threaten rational scientific inquiry" underpinning environmental and other issues, and Prince Charles (2000) has worried about how religiously based notions of stewardship for the earth have been "smothered by almost impenetrable layers of scientific rationalism." Is the dual invocation of science and religion a zero-sum affair when the two are mixed, where hard, cold science disenchants nature, or wooly-headed inspiration trumps logic? The Ehrlichs and Prince Charles are not alone in their suspicions to this effect.

Whether one mixes or separates, whether one includes or excludes, all these proposed settlements to the question of science and religion have direct implications for how we know and care about biophysical nature. All are settlements in that they fix science, religion, and nature once and for all: Nature as revealed only by science, nature as sacred, nature as a confluence of concern for the great traditions of science and religion. But all these settlements ignore the great contradictions of our time, the larger context of the old-growth issue. Here we are, a hyperconsumptive culture living in lumber-hungry homes twice the size they were in the 1950s, keen on the cult of the market that overlooks its impacts on all landscapes save those such as old growth that resonate deeply with science and spirituality.

Somehow it all makes sense when we focus on old-growth protection, but when we move further along this larger circle it makes no sense whatsoever.

Forging a New Idea of Nature

This is why we need a new approach to nature and why this new nature will, among other qualities, be built on a new sense of science and religion. I propose we admit that nature is wonderfully complex and ultimately irresolvable, with an epistemological and moral dynamism revealed in part by the utter ambivalence of science and religion as joint authoritative voices on nature. The term *ambivalence* is from the psychological literature of the early twentieth century. It means, literally, "both strengths." There is a strength to scientific knowledge, a strength to religious insight. But when you mix them you become ambivalent. Science and religion present us with a set of paradoxes about nature given their ambivalent guidance, and paradox may be a good thing to the extent that it instills a proper sense of humility on all sides of debates over saving nature (Proctor, 1998b, 2001).

What of this new nature as it applies to old-growth forests? It may lead us to be more tolerant of the differences one senses when reading the chapters that constitute this volume or even to a greater extent when listening to all the voices offering their opinion on management of old-growth forests. These differences may be, contrary to our wishes, as conceptually irresolvable as the mix of biochemistry texts and country churches that proved so important to my upbringing. No one has the final word, as much as she or he may wish to do so. Old-growth forests do offer a measure of sacredness as well as carbon sequestration, but the sum of the two is not tidy and does not offer a clear resolution to policy battles over old-growth forests, no matter how much we wish it would.

Yes, ultimately we must agree on policy affecting the public old-growth forests we share, but if we are looking for the vast assemblage of facts and values about old-growth forests to steer us unequivocally toward a policy resolution, we are looking in vain. An ambivalent nature is anything but obvious, and anyone who claims otherwise in the case of old-growth policy is speaking with the very hubris an ambivalent nature belies.

Speaking of a new conception of nature thus involves speaking with a bit of a stutter in our authority. Old-growth forests become part of a humbling experience in that we cannot definitively trace a conclusive circle around them. They are certainly different, ecologically and spiritually, form other landscapes. Yet we would speak of the need to bolster protection of old-growth forests, or the need to remove protection from old-growth forests, with equal hesitancy, because we would realize that we base our argument around an ambivalent set of authorities.

This insecure terrain is not for the faint of heart. It may take genuine engagement among citizens, scientists, interest groups, and policymakers to rediscover a shared ground common enough to lead to lasting policy. But, to the extent that old-growth forests served as the icon of an ancient, settled nature at the close of the twentieth century, they may indeed help us usher in a new ambivalent nature in the twenty-first. If we indeed find, and learn to care for, this new nature in old-growth forests, we will surely find and care for it elsewhere.

LITERATURE CITED

Albanese, C. L. 1990. *Nature religion in America: From the Algonkian Indians to the New Age, Chicago history of American religion*. Chicago: University of Chicago Press.
———. 2002. *Reconsidering nature religion*. Harrisburg, PA: Trinity Press International.
Carson, R. 1962. *Silent spring*. Boston: Houghton-Mifflin.
Ehrlich, P. R., and A. H. Ehrlich. 1996. *Betrayal of science and reason: How anti-environmental rhetoric threatens our future*. Washington, D.C.: Island Press.
Einstein, A. 1954. *Ideas and opinions*, translated by S. Bargmann. New York: Crown Publishers, Inc.

Gould, S. J. 1999. *Rocks of ages: Science and religion in the fullness of life*. New York: Ballantine Publishing Group.

Mission to Washington. 1996. Declaration of the Mission to Washington. In *This sacred earth: Religion, nature, environment*, edited by R. S. Gottlieb. New York: Routledge.

National Religious Partnership for the Environment. 2004. *Earth's climate embraces us all: A plea from religion and science for action on global climate change*. Amherst, MA. http://www.nrpe.org. [accessed 9/12/04].

Pollan, M. 1991. *Second nature: A gardener's education*. New York: Atlantic Monthly Press.

Prince Charles. April/May 2000. *Millennium Reith Lecture*. http://news.bbc.co.uk/hi/english/static/events/reith_2000/lecture6.stm. [accessed 1/1/06].

Proctor, J. D. 1998a. Geography, paradox, and environmental ethics. *Progress in Human Geography* 22(2):234–55.

———. 1998b. The spotted owl and the contested moral landscape of the Pacific Northwest. In *Animal geographies*, edited by J. Emel and J. Wolch. London: Verso Press.

———. 1999. A moral earth: Facts and values in global environmental change. In *Geography and ethics: Journeys in a moral terrain*, edited by J. D. Proctor and D. M. Smith. London: Routledge Press.

———. 2001. Solid rock and shifting sands: The moral paradox of saving a socially constructed nature. In *Social nature: Theory, practice and politics*, edited by N. Castree and B. Braun. Oxford, U.K.: Blackwell Publishers Ltd.

———. 2005. Introduction: Rethinking science and religion. In *Science, religion, and the human experience*, edited by J. D. Proctor. New York: Oxford University Press.

Proctor, J. D., and E. Berry. 2005. Social science on religion and nature. In *Encyclopedia of religion and nature*, edited by B. Taylor. New York: Continuum International.

White, L., Jr. 1967. The historic roots of our ecologic crisis. *Science* 155(March): 1203–07.

Williams, M. 1989. *Americans and their forests: A historical geography*. Cambridge, U.K.: Cambridge University Press.

Chapter 10

Common Sense Versus Symbolism: The Case for Public Involvement in the Old-Growth Debate

BRENT S. STEEL

After reviewing the aftermath of Oregon's 2002 Biscuit Fire in the Kalmi-opsis Wilderness, President George W. Bush used the symbol of old-growth forests in a speech to encourage the public and Congress to support his Healthy Forests Initiative (which eventually became the Healthy Forest Restoration Act of 2003). At the core of this initiative were two proposals that would actually lead to increased harvest of trees, potentially including old growth: (i) a new streamlined decision-making process to limit appeals to logging plans and (ii) the need to thin forests to avoid such fires in the future. President Bush argued that, if enacted by Congress, the Healthy Forests Initiative would help "save" old-growth trees from catastrophic fires through increased harvest of old and newer growth trees. However, opponents saw the initiative as nothing more than Orwellian doublespeak, promising to save old-growth forests but actually opening the door for cutting of such currently protected forests. Although scientists have debated and continue to debate the wisdom of thinning forests in differing conditions, the symbolic use of the descriptor *the oldest trees* by President Bush to build support for his initiative was no accident. Old-growth trees and forests have become a powerful icon of nature in the Pacific Northwest, and all interests involved in the forest management policy debate use these icons as symbolic cover for their own policy preferences.

The use of symbols has been an integral part of the American political landscape since the boom in environmental legislation of the 1960s and 1970s, when environmental interest groups and their industry opponents discovered their power in the realm of natural resource management. These divergent groups have typically found success by simplifying a complex issue and selling it hard in the political marketplace, and the fate of old-growth forests has been no exception. The key elements in the social conflict over old-growth forests include a visceral sense of beauty and eternity *versus* a sense of entitlement to wealth and an expanding and bewilderingly complex scientific knowledge base *versus* a utilitarian view of natural resources. What the debate about old growth perhaps best illustrates is just how poorly special interests serve the public when the issue is complex.

Wealth, Beauty, and Special Interests

In his book, *The Final Forest: The Battle for the Last Great Trees of the Pacific Northwest,* William Dietrich argued that forests are part of our common heritage—regardless of political jurisdiction or ownership—that provide both "wealth and beauty" the "twin desires of the human heart" (1992: 25). These twin desires have led to and will continue to lead to conflict over the management of old-growth forests.

I believe a key ingredient in resolving some of the conflict resides in making forest policymakers accountable to the considered will of the people instead of the whims of highly polarized, take-no-prisoners interest groups and industry. However, this is increasingly difficult due to the scientific, technical, and political complexity of such issues, which in turn poses serious challenges for the effective participation of citizens in the democratic process. For citizens to monitor policymakers effectively in democratic societies, they need to be informed consumers of relevant scientific research and the policy options suggested by those findings: Increasing public understanding of environmental problems can build capacity for solving those problems. As President James Madison once warned, "A people that mean to be their own governors must arm themselves with the power that knowledge gives."

However, the critical gaps between the need for policy-relevant knowledge, the generally poor level of public understanding of many public policy issues, and the genuine intent of many political maneuvers have led some commentators to proclaim the existence of a "legitimacy crisis," leading to low levels of participation or even vacuous input when it is ill-informed. The array of information available on television and the Internet, and its highly variable levels of credibility, have not helped. In fact, the electronic media have made it much easier for special interests to use symbols and icons, such as old-growth trees, to obfuscate the policy process. They now package and present verbal symbols (e.g., *thinning* instead of *cutting*) and visual symbols (e.g., the adorable creatures that frequent forests, such as bunnies and birds) to generate attention and support. In this way, the complexity of policy issues is reduced while playing on the emotions of the public.

From Pluralist Paralysis to Social Movement

Without engaging the general public in a meaningful way, the cast of usual suspects — industry, other special interest groups, politicians, etc. — will continue to frame the old-growth debate and perpetuate its conflict through symbolic manipulation. Often "public" participation processes in natural resource management are dominated by interest group representatives and industry and barely even involve the general public. Although their participation in the debate gives the impression of "democracy," interest groups and industry have their own agendas to pursue, often encouraging polarization, conflict, and stalemate, as typified by the Northwest Forest Plan (a plan in name only). Political scientists call this *pluralist paralysis*.

The key to escaping pluralist paralysis is to design forest management institutions and processes in a manner that informs and encourages meaningful public participation. Some have even suggested that many interest groups involved in natural resource policy issues should abandon "lobbying" the government and focus their time and resources on educating and mobilizing the public into a broad-based *social movement* (Gonzalez, 2005). Social movements are grassroots movements sufficiently broad-based to emphasize a collective identity, values, and lifestyles. They encapsulate a wide range and large number of organizations and individuals united for a particular cause. The argument is that such broad unity would force government to broaden and democratize the management of the environment, including the old-growth forest issue.

Research concerning general public orientations toward this debate suggests a less-polarized and broader-based perspective than the ones evident among interest groups currently involved in the old-growth debate, indicating that the conditions necessary for a social movement are already present. For example, random sample surveys of Oregon households conducted by researchers at Oregon State University (Shindler et al., 1993) in 1991 and 2004 indicated that many people would like to balance the "twin desires" of wealth and beauty in terms of federal forest management, with a slight bias toward beauty (table 10.1). In the 1991 survey, seventy-two percent of respondents can be found in the middle of the socioeconomic versus environmental priority scale (responses three, four, and five) compared to seventy-four percent in the 2004 survey. It may come as a surprise to many that the perspective of Oregonians on this issue has remained very stable between the time when the old-growth controversy was first hitting its peak in 1991 and the most recent survey in 2004. A balance of wealth and beauty is consistent with many of the management goals promulgated by the National Research Council's 2000 report, *Environmental Issues in Pacific Northwest Forest Management*.

TABLE 10.1. Oregon public preferences concerning economic and environmental trade-offs, 1991 and 2004.

Question: *Federal forest management may require difficult trade-offs between restoring environmental conditions (e.g., wildlife, old-growth forests) and socioeconomic considerations (e.g., employment, tax revenues). Where would you locate yourself on the following scale concerning this issue?*

		Scale	% 1991	% 2004
The highest priority should be given to environmental conditions, even if there are negative socioeconomic consequences.	‡	1	10	10
		2	12	13
		3	14	14
Environmental and socioeconomic factors should be given equal priority.	‡	4	45	48
		5	13	12
		6	4	2
The highest priority should be given to socioeconomic considerations, even if there are negative environmental consequences.	‡	7	2	1

Source: Program for Governmental Research and Education, Oregon State University.
NOTE: Sample sizes: 1991 = 876 respondents; 2004 = 1,512 respondents.

Muddying the Debate with Intention

However, additional surveys of interest groups and industry representatives in Oregon conducted by Oregon State University researchers found extremely polarized and fragmented responses to the question of balance in management goals. For example, a 1996 study of Oregon interest groups and industry identified several hundred interest groups attempting to influence the policy process concerning the management of public forests and rangelands (Steel et al., 1996). The types of groups formally involved in the Oregon natural resource and environmental policy process included the following: environmental protection and conservation groups (e.g., Portland Audubon Society), intensive recreation groups (e.g., Pacific Northwest Four Wheel Drive Association), passive recreation groups (e.g., Mazamas, a Portland-based mountain climbing group), industry-related groups (e.g., Western Forestry and Conservation Association), and industry (e.g., Oregon Forest Industries Council). Another set of important actors in this policy milieu is scientists who, depending on their area of expertise, source of funding, and general environmental worldview, can also provide conflicting and polarized policy prescriptions. A case in point would be the recent furor in Oregon State University's College of Forestry, where faculty clashed publicly over the wisdom of salvage logging, revealing rifts that had complicated connections to worldview, funding, age, and expertise. From such conflicts emerge public perceptions of science and scientists as being driven by funding sources or ideology, creating an atmosphere of cynicism and distrust, and thus further complicating their ability to inform the public.

The sheer number, diversity and intensity of interests, and high level of political activity at virtually every stage of the policy process dilutes the impact of the general public in the old-growth policy process. Indeed, one could argue that it further complicates already conflicting, confusing, and biased perspectives. The scientific and technical complexity of forestry issues in general, and of old growth more specifically, poses serious challenges for the effective participation of citizens in the policy process. Although citizens have the right to monitor and influence policymakers in democratic societies, they cannot do so effectively if they are left in the dark on (i) real policy options, (ii) the underlying assumptions behind those policy options, and (iii) the major findings of relevant scientific research.

A further complicating factor discussed above, but not always perceived or understood by the public, is the role of what political scientists call *symbolic politics*. This occurs when speeches or gestures are made—such as the speech by President Bush quoted in the introduction—that appear to address the problem at hand, people applaud, everyone goes home, and nothing further is done. In the case of old-growth forests, two examples illustrate this: (i) the government notes that some old-growth forests are still open for logging (to appease industry), but no logging is done (to appease environmentalists), and (ii) environmentalists protest logging on the basis of protecting "old-growth" forests, even when the forests are not particularly old or the logging is designed for ecological restoration that really *could* preserve old-growth stands.

How and What Does the Public "Know"?

Many social scientists have argued that public knowledge is central to democratic policymaking processes. Without public knowledge, there may be little understanding of problems, no public awareness, and consequently no democratic policy process. This line of argument reflects the assumption that knowledge is essential to understanding the implications of policy alternatives and that knowledge can be a source of substantial political influence for citizens to take political action. It follows, perhaps, that an ill-informed public—one that does not grasp the complexity of particular environmental issues—can be more easily manipulated by elites and special interests using symbolic images and language. Symbolic solutions can then follow.

Given the importance of policy-relevant knowledge to how the policy process takes shape in democratic societies, what levels of knowledge concerning old-growth ecosystems issues are apparent among citizens? Although the public wants an environmentally balanced forest policy, their self-assessed knowledge of terms used by scientists and managers in the old-growth policy debate is mixed (table 10.2). Survey data indicate that there is general understanding of such terms as *ecosystem* (eighty-eight percent familiar) and *old-growth forest* (eighty-four percent familiar), but there is less understanding of more technical terms such as *keystone species* (twelve percent familiar), and *silviculture* (twelve percent familiar). What is particularly interesting is that, although eighty-four percent say they know what *old-growth forest* means, only sixty-five percent claim to know *ancient forest* and twenty-two percent *late-successional forests*, all terms used to describe the same type of ecosystem. This suggests that the use of multiple terms may lead to some public confusion. Another important item where there was a relatively low level of self-assessed familiarity was with the *Northwest Forest Plan* (forty-two percent familiar). This level of public familiarity would have to increase before effective public participation could productively take place.

TABLE 10.2. Oregon public's familiarity with forestry terms 2004.

Term familiarity: *For the following terms, please indicate if you 1 = know what the term means, 2 = have heard of the term but don't know its meaning, or 3 = have not heard of the term at all.*

Term	% Say "Know" Term
Ecosystem	88
Old-growth forest	84
Managed forest	69
Ancient forest	65
Biodiversity	61
Conifers	58
Northwest Forest Plan	42
Late-successional forests	22
Keystone species	12
Silviculture	12
Multiple canopy layers	5
Secondary forest	3
Disturbance regime	2
Vegetative succession	2
Tree decadence	1
Terrestrial vertebrate species	1

Source: Master of Public Policy Program, Oregon State University.
NOTE: Sample size: 1,512 respondents.

Although these results are not conclusive evidence of low levels of public knowledge concerning the old-growth debate, they do suggest a need for care in the language we use to engage the general public as well as a need for public education outreach efforts. Communications directed to the general public are important not only because they may influence public opinion and, therefore, have an impact on public policy but also because they are potentially effective in inducing individuals to engage in behavior that can lessen the destructive impact of humans on the environment. In fact, most research by political scientists shows a strong positive correlation between policy-relevant knowledge and meaningful and effective political participation. However, the use and continued use of verbal and visual symbols by special interests to play on emotions of the public will make outreach efforts extremely difficult.

How old growth is defined is crucial to the framing of the policy

Table 10.3. Oregon public's definition of "old-growth" forest 2004.

If you indicated that you are familiar with the term "old growth," please indicate below how old trees should be to be considered old growth:

Age	%
50 years	3
51–100 years	6
101–150 years	8
151–200 years	19
201–250 years	23
Over 250 years	14
Don't know	27

Source: Master of Public Policy Program, Oregon State University.
NOTE: Sample size: 1,253.

debate. The National Research Council's report *Environmental Issues in Pacific Northwest Forest Management* stated that "no single definition of old growth is appropriate"(2000: 46). Although there are numerous characteristics of old-growth forests to consider, we asked the public to indicate how old trees should be before they can be considered "old growth." Obviously, definitions of old growth can be politically charged and symbolic, and it is useful to see how this issue is perceived by the public.

The survey data presented in table 10.3 indicate that twenty-seven percent of the public does not know how to define old growth. However, several of these same respondents replied that they "know it (old growth) when they see it." The majority of respondents (fifty-six percent) believed that it is more than 150 years old. It could be that "beauty" leads to the public's definition of old growth instead of scientific or other definitions, thereby complicating this issue for managers and policymakers. However, because there is no scientific consensus on this matter, one could hardly expect the public to have one either.

The data displayed in table 10.4 examine general public support for commodity-based versus ecosystem-based forest management policies in 1991 and 2004. What is interesting is the relative stability of support during this period of time for most management approaches across the spectrum. In general, however, people have been and are still concerned about wildlife, wilderness, and old-growth forests. In fact, support for the statement "Greater efforts should be made to protect the remaining old-growth

TABLE 10.4. Oregon public support for federal forest management policies, 1991 and 2004.

Please indicate your level of agreement or disagreement with the following statements concerning the management of federal forest lands (i.e., USDA Forest Service and Bureau of Land Management).

	% Strongly agree		% Agree		% Disagree		% Strongly disagree	
	1991	2004	1991	2004	1991	2004	1991	2004
Commodity-based orientation:								
The economic vitality of local communities should be given the highest priority when making federal forest decisions.	20	18	26	24	27	26	17	15
Some existing wilderness areas should be opened to logging.	11	8	21	16	16	23	35	38
Endangered species laws should be set aside to preserve timber jobs.	13	10	24	21	17	21	31	35
Federal forest management should emphasize timber and lumber jobs.	16	14	16	15	21	22	18	19
Ecosystem-based orientation:								
Clearcutting should be banned on federal forest land.	35	38	22	25	18	15	12	9
More wilderness areas should be established on federal forest lands.	25	29	21	22	12	13	18	12
Greater efforts should be made to protect the remaining "old-growth" forests.	35	41	16	18	17	14	15	12
Greater efforts should be given to wildlife on federal forest lands.	30	33	25	27	16	15	9	9

NOTE: Sample sizes: 1991 = 876 respondents; 2004 = 1,512 respondents. "Uncertain" responses not shown in table.

forests" increased from fifty-one percent agreeing or strongly agreeing with this statement in 1991 to fifty-nine percent in 2004.

These data show that the Oregon public would like a balanced forest policy that provides for wealth and beauty but does not sacrifice old growth, wilderness, and wildlife protection. Of course, this situation is similar to other areas of public policy in which the public likes "having its cake and eating it too." However, there is some degree of polarization evident in these results with forty-two percent of respondents in 2004 agreeing that the "economic vitality" of local communities should be given highest priority compared with forty-one percent disagreeing with this statement. Taken with the results displayed in table 10.1, it appears that there is concern for timber-dependent rural economies only if old growth, wilderness, and endangered species are not harmed. The results here and in other public opinion studies show support for timber harvest on public lands if harvesting does not jeopardize "beauty."

Looking for Answers

Clearly any solution to the debate about old growth will involve painful trade-offs and contentious argument. For example, there are some significant differences between rural and urban Oregonians for several of these issues, such as endangered species. However, for most citizens these differences are not as pronounced as the positions of many interest groups and industry currently involved in the process. Moving from interest group– and industry-driven policy (a form of symbolic democracy) to a process that truly embraces and involves the general public may not only moderate this debate between wealth and beauty but also bring about a more democratic, common sense, and therefore legitimate result.

Although there is no assurance that meaningful and informed public participation processes will lead to consensus, it is clear that some noteworthy successes at collaborative and participatory public environmental policy processes have occurred in many locations. Examples of successful (and unsuccessful) efforts can be found in Ed Weber's work on collaborative resource management (2003) and a recent study reported in *High Country News* (Durbin, 2006). When collaboration is pursued with integrity, the number of people who cannot be manipulated by symbolic verbal and visual gestures and utterances increases rapidly and productively (Koontz, 2006).

It is also clear that "civic journalism" (Dahlgren and Sparks, 1991) can do much to promote the type of public discussion and deliberation that occasions a "coming to public judgment" (Yankelovich, 1991). Currently, in our "sound bite" political world, the duplicity of politicians and media in using symbols instead of substance merely confuses a truly engaging discussion of such complex policy issues as old growth. The challenge we face is that until we have meaningful public discourse, federal government agencies retaining control of public lands in the West will continue to engage in battles about their policies, and even about their very authority. As a result, they too will need to promote public participation while supporting the efforts of responsible journalists to educate the public on old growth and other environmental issues. Federal agencies will also have to rebuild public trust by providing honest and useful information to the public and media and avoiding symbolic politics themselves.

At its best, civic journalism can help drive a stake into the heart of special interests and symbolic politics, a powerful combination that has proven its worth in muddying the waters of public knowledge and stopping intelligent progress in the thoughtful management of natural resources. Old-growth forests, in particular, represent the kind of complex environmental and political issue that requires a changed view of how the democratic process can best serve its constituents.

LITERATURE CITED

Dahlgren, P., and C. Sparks. 1991. *Communication and citizenship: Journalism and the public sphere in the new media age.* New York: Routledge.

Dietrich, W. 1992. *The final forest: The battle for the last great trees of the Pacific Northwest.* New York: Simon & Schuster.

Durbin, K. 2006. The war on wildlife. *High Country News* 38:8–13, 19.

Gonzalez, G. 2005. *The politics of air pollution: Urban growth, ecological modernization, and symbolic inclusion.* New York: SUNY Press.

Koontz, T. 2006. Collaboration for sustainability? A framework for analyzing government impacts in collaborative–environmental management. *Sustainability: Science, Practice, and Policy* 2(1):15–24.

National Research Council. 2000. *Environmental issues in Pacific Northwest forest management.* Washington, D.C.: National Academy Press.

Shindler, B., P. List, and B. S. Steel. 1993. Managing federal forests: Public attitudes in Oregon and nationwide. *Journal of Forestry* 91:36–42.

Steel, B. S., J. C. Pierce, and N. P. Lovrich. 1996. Resources and strategies of interest groups and industry representatives in federal forest policy. *Social Science Journal* 33:401–21.

Weber, E. 2003. *Bringing society back in: Grassroots ecosystem management, accountability, and sustainable communities.* Cambridge, MA: MIT Press.

Yankelovich, D. 1991. *Coming to public judgment: Making democracy work in a complex world.* Syracuse, NY: Syracuse University Press.

PART III

Value, Conflicts, and a Path Toward Resolution

The old-growth controversy is a classic case of conflicting ideas around human interactions with the natural world, long predating recorded history. Always at the heart of such interactions lie fundamentally different social, philosophical, and spiritual perspectives on our relationships with nature. This section highlights different ways of expressing how humans have related to old growth in the immediate past and how those relationships affect our current and future old-growth policies. We have alternated different perspectives to give a sense of the debate that went on and still goes on today.

Kerr, a leading environmentalist and old-growth advocate, relates how environmentalists latched on to early scientific findings to help make their case and how successful they were. Now that the old-growth battle is over, he sees a role for the forest industry in restoring diversity and resistance to fire. Sohn, a private forest owner and state policy advisor, reflects on how some industrial forest owners view the issue (old growth is nice to have but not their job) and argues that emotions have gotten in the way of a mutually respectful dialog. Brown, an environmentalist, describes how views of "natural forests" as places where humans don't manage are challenged in fire-prone forests with a long history of fire suppression. Mickey, an industry association representative, challenges the scientific basis of our understanding of old growth, points out ambiguities in definitions, and marvels at how people can proclaim facts and positions about a concept that he feels is vague and ill-defined. He concludes, however, with a call for action to manage for the production of old growth. Moore, a nature writer and philosopher, identifies the instrumental spiritual values of old growth that have utility to humans but wonders if they are just an expression of the deeper intrinsic values that we feel but cannot express when we experience an old-growth forest. Finally, we conclude this section with some observations by Wondolleck, an academic who has specialized in conflict resolution problems. She provides lessons learned from studies of intractable conflicts, including the maxim that issues involving fundamental value differences are not suited to collaborative resolution.

Chapter 11

Starting the Fight and Finishing the Job

ANDY KERR

It wasn't until the 1970s that scientists started taking a closer look at old-growth forests in the Pacific Northwest. Until then these forests had been typecast as "biological deserts" to justify their replacement by timber plantations. Although they did not fully realize it, the pioneer scientists poking around in these uncharted ecosystems were on the verge of opening a magnificent treasure trove of discovery, one that would affect the region's development for many decades to come.

In the late 1970s, I attended a meeting with U.S. Forest Service research scientists on the Oregon coast, where I ended up in the bar at the Inn at Otter Crest. On my left was Dr. Jerry Franklin, even then the preeminent authority on Pacific Northwest forests. On my right was Dr. Jack Ward Thomas, even then the preeminent authority on Rocky Mountain elk. Both had had more than one regional forester try to get them fired for their science. Fortunately, they were in the Forest Service's research branch, and those who wanted them fired were on the agency's National Forest System side. They went on to become the two most powerful scientists in public forest policy, figuratively being carried to greatness by the talons of the northern spotted owl.

The presence of alcohol and the passage of time have caused me to forget most of the bar talk. However, I remember Jerry being

129

pessimistic about the timber juggernaut's ever being reined in. At the time, two square miles of Oregon old-growth forest was clearcut each week. When Dr. Old-growth was so down, I could but quietly cry in my beer. I was doing so, when Dr. Wildlife loudly weighed in with optimism that things could and would change. Awed by both, I said nothing. Jerry, ever the skeptical and inquiring scientist, asked Jack why he thought things would change. Jack boomed out (concurrently punching my right shoulder with enough force that I had to use my left arm to counterbalance), "Because sons-of-bitches like him are going to cause trouble."

Who was I to argue with science? Actually, at the time there wasn't much published science with which to support sons-of-bitches' arguments, but that barrier was soon to be breached by a strategic leak. A few years before it was published in 1981, a Forest Service research branch draft report was leaked to the Oregon Natural Resources Council (ONRC). Conservationists feared that political pressures from Big Timber would prevent publication of the report, *Ecological Characteristics of Old-Growth Douglas-Fir Forests* (Franklin et al., 1981). As a result, enough draft copies of the report were passed hand-to-hand that it became a classic even before publication.

This fifty-two-page booklet energized the emerging debate over old forests. On the one-page summary of sixteen points, the eight authors exploded the major myths surrounding old-growth forests. They are not biological deserts. Dead trees—both standing and downed—are ecologically critical. Productivity is very high. Structure and function are very complex. Natural stream function depends upon old trees.

As is often the case when the emperor is wearing no clothes, speaking up has powerful political repercussions.

Concurrently, other scientists were learning about what would become the ultimate flagship and indicator species for the old-growth forest. The first northern spotted owl I saw was in a cage in Eric Forsman's (Dr. Spotted Owl) (see box 4.1) backyard in Corvallis. I recall no epiphany from this interspecies eyeballing, but it must have had some effect.

Initially, conservationists feared petitioning to list the species under the Endangered Species Act. It was a time when Oregon U.S. Senator Mark Hatfield—and other powerful politicians with sawdust coursing in their veins—would readily change the law if conservationists were successful in using it to limit old-growth logging. Eventually conservationists realized that the law is only meaningful if enforced, and so began a barrage of lawsuits citing a panoply of federal statutes and regulations.

First the Words . . .

In the early 1980s, as ONRC was bringing the first administrative appeals of timber sales in spotted owl habitat, we debated calling these magnificent natural forests anything but *old growth*. We had our reasons.

First, foresters then defined *old growth* as any tree older than silvicultural rotation age, or culmination of mean annual increment the point at which the maximum growth rate of a tree begins to slow. For Douglas-fir, this was around eighty years of age. To maximize wood volume, a forester would then clearcut and replant.

Second, the timber industry defined *old growth* as any tree older than the economic rotation age (age at which harvest will generate maximum revenue). Because time is money and even young trees have value once they reach a certain size, trees more than twenty-five to forty years of age were "old growth" to a capitalist. To maximize shareholder value, an accountant would then clearcut and replant (but only because the law required such; growing trees is not particularly profitable).

Third, scientists didn't have a common definition of *old growth*. They still don't. In hindsight, this is understandable because old-growth forests don't just become this way overnight, and characteristics vary by species, site, and other factors. Supreme Court Justice Potter Stewart said it of pornography, but he could have applied the thought to old-growth forests as well: "I shall not today attempt further to define the kinds of material I understand to be embraced . . . [b]ut I know it when I see it."

Finally, America is a youth-worshiping society, and *old* is generally not perceived to be desirable. In addition, *growth* as a noun is something one has a surgeon remove.

I favored calling it *primeval* forest, recalling Henry Wadsworth Longfellow's *Evangeline: A Tale of Acadie* (1847):

> *THIS is the forest primeval. The murmuring pines and the hemlocks,*
> *Bearded with moss, and in garments green, indistinct in the twilight,*
> *Stand like Druids of eld, with voices sad and prophetic,*
> *Stand like harpers hoar, with beards that rest on their bosoms.*

ONRC Executive director James Monteith argued that *primeval* was an archaic word more likely to be interpreted in this modern era as "prime evil." He favored *ancient* forest ("having had an existence of many years, having the qualities of age or long existence, venerable") and carried the argument by noting that "ancient" had but two syllables and seven characters.

Thereafter, we started referring to ancient forest. Within two weeks, "ancient forest" was in a headline of a story in the *Eugene Register-Guard*. The timber industry howled, citing the fourth dictionary definition of "ancient" ("prior to the fall of the Roman Empire in 476 AD") and noting that old growth isn't that old.

Actually, some of it is.

Conservationists were about to confirm their suspicions that controversy could be the forest's best friend. As Frederick Douglass noted in the nineteenth century, "If there is no struggle there is no progress." Like it or not, social progress comes from social tension—and most change comes at funerals.

Besides "branding" the ecosystem for political marketing, attention had to be drawn in other ways besides litigation. One successful strategy was nonviolent civil disobedience—specifically, tree sitting. Although many would plot tree sitters as dwellers on one edge of the human bell curve, tree sits helped bring attention to ancient forests and generate media coverage. Initially, the regional media were missing the story—precisely because it was local. Before the old-growth issued "flamed" publicly, I toured a British reporter through some old-growth forests. It became a big local news story that a reporter from the *Times* of London was covering a Pacific Northwest issue.

The Oregon congressional delegation was stunned when I was quoted in *Time* magazine as saying that expecting them to deal rationally with the end of ancient forest logging in the 1980s was like expecting the Mississippi delegation to deal rationally with the end of segregation in the 1950s (Seideman, 1990). Senator Hatfield was livid and yelled at the *Time* editors (the magazine later used the quote again). For the next decade, Congressman Ron Wyden (later to succeed Hatfield in the Senate) referred to the quote every time he saw me. Both were so stunned because both were active for civil rights. However, both were politicians of their time and place, not unlike Strom Thurmond and George Wallace.

Then the Science . . .

The role of science in informing, guiding, controlling (and yes, sometimes confusing) the debate about the future of Pacific Northwest forests has been remarkable. More than once, I found that the latest scientific recommendations went beyond what conservationists had been advocating.

Science played such a crucial role in the old-growth forest logging controversy because the controversy was a politically unsolvable issue. The revelations about the warm-blooded and extremely mediagenic northern spotted owl's dependence upon old-growth forests set up the political paradox. As more science came out about the habits and habitat needs of the species, the politicians didn't know what to do. Playing to Big Timber's arrogance and paranoia (denial is not just a river in Egypt) and also playing for time, the political response was often to call for more science ("study"). Out it would come. When the truth about spotted owls could no longer be denied, the *timber-industrial complex* (timber industry, timber bureaucracy, and timber state politicians) said that old-growth dependency may be true for one species but certainly not others. Then came concern about the marbled murrelet, Pacific salmon, and potentially thousands of other species (not the least of which is the Malone jumping slug).

In 1991, congressional staffers convened a meeting of research scientists and agency bureaucrats at the Inn at the Quay in Vancouver, Washington. Although not invited, conservationists rented the closest room and stocked it with booze to lure the thirsty attendees after their meeting. Yet another scientific exercise was commissioned during the meeting, and later I was lobbying Jack Ward Thomas and Jerry Franklin (this time *standing* with our drinks) to include Pacific salmon stocks in the analysis. Jerry—understandably feeling daunted by the task and timetable just handed him—resisted, but Jack reflected a moment and loudly said, "We're going to do it." I am not suggesting that incorporating Pacific salmon into the political cocktail that started with the spotted owl was due to my lobbying prowess or a drink loosening Jack's tongue, although alcohol may have freed both our minds a bit. It was simply inevitable that the public debate over these two iconic organisms would be combined.

Eventually, the timber industry wised up and quit trying to fight the science of old-growth forests. They have since diverted the context of the debate to focus on what to do after natural disturbance of a forest (fire, windthrow, insects, disease, etc.), where the science wasn't as well developed. In addition, the public has been brainwashed by Smokey Bear, and most think that a singed forest (poor Bambi) is a lost forest that might as well be clearcut and planted. The last best chance for the timber industry to log federal forest lands is by exploiting the public's unbounded ignorance of wildfire. The public loves old trees, scenic forests, healthy watersheds, and roadless areas. The public does not love burned forests.

Yet.

As political controversy deepened, so did the science of old-growth forests. Each new scientific report noted that it was necessary to conserve and restore even more land to maintain a functioning forest ecosystem, both across the landscape and through time.

Today, the debate over the definition(s) of old-growth forests has been won by scientists and accepted by the public, although scientists themselves continue to dicker over the details. Meanwhile, Oregon Wild research shows that both "old-growth forest" and "ancient forest" poll equally well.

And Finally, the Religion

Although science drove the Pacific Northwest forest debate publicly, not readily acknowledged was the religious dimension. There is no doubt that religion was an undercurrent of the economics versus ecology debate, although all sides generally avoided any public mention of it.

Although there were exceptions to the rule, those who fervently advocated for ancient forests were motivated not only by the science but also by spiritual and moral components. These components were expressed in various ways: Christian stewardship (i.e., we don't have a right to destroy what God created), worship of nature (some people do hug trees), and/or worship of science (in which science coincided with core beliefs).

Equally the case, those who fervently advocated against old-growth forests were motivated not only by greed but also by spiritual and moral components. These components were expressed in various ways: Christian stewardship (God gave us these forests to use and improve), aversion to nature (the wild is to be tamed for the good of all humankind), and aversion to science (because science conflicted with core beliefs). An acceptance or rejection of the science of evolution is a good—but, of course, not perfect—predictor of the stance one takes on leaving forests standing.

Today, the debate over old-growth forests is finished. Polls tell us that vast majorities of the public don't want any more old-growth forest logged. They don't even want any "mature" forest logged—80- to 149-year-old trees. Federal forest logging levels everywhere have fallen dramatically. Most of the timber industry has downsized and resized to survive mostly on smaller logs from private lands. The number of mills still operating in the Pacific Northwest that are capable of even milling a very-large-diameter log can be counted on two hands (no need for feet).

President Bill Clinton was a big tree between a couple of bushes in more ways than one. The spotted owl hit the fan during the term of George H.W. Bush, and the timber industry counterattacked during the terms of George W. Bush. In between, Clinton, who was a very successful politician and also a policy wonk, solved the Pacific Northwest old-growth forest issue—politically, although not ecologically. Yet, as in the Battle of the Bulge, the timber industry, led by George W. Bush's undersecretary of agriculture Mark Rey, counterattacked with great force, seeking to undo the Northwest Forest Plan. Nonetheless the fall of Big Timber—perhaps just like that of Berlin—was inevitable.

The challenge for conservationists was to provoke a political crisis commensurate with the ecological crisis. By all measures, they achieved this. The dramatic change in public policy toward Pacific Northwest old-growth forests is unprecedented in conservation politics and most other kinds of politics as well. Northwest public timber-cutting levels plummeted eighty to ninety percent in the course of this relatively civil war for forests. The political half of my brain is very pleased with the progress, but the ecological side knows that while the worst bleeding has been staunched, the patient is still losing blood and is quite weak.

So Where Are We Going?

Enough reminiscing about the past thirty years. It's time to speculate on the next three decades of public forest and private timber management in the Pacific Northwest. And if I suffer from wishful thinking in this endeavor, it is no more or less than I always have.

- *More people.* The region will become more urbanized. In the early 1990s, a poll revealed that a vast majority of Oregonians didn't even know anyone who worked in the timber industry, never mind did they work in it themselves. This will increasingly become the case as the timber industry continues to shrink in economic and social importance, in both relative and absolute terms.
- *Competing uses for biomass.* The economics of forest biomass will change. As the United States comes to its senses on global warming, it won't be just about reducing fossil fuel emissions but about transferring excess atmospheric carbon back to the biosphere. Growing—and not logging—forests is the most proven form of carbon sequestration. In time, the carbon industry will outcompete the timber industry in acquiring biomass.

- *Naturally occurring young forests.* More rare than old-growth forests, young forests that naturally developed after disturbance will be greatly valued in the future (after an initial period of controversy over salvage logging) because they are extremely important biologically, having the legacy of the previous old forest and the productivity of the young forest. Such forests are rare because of decades of systematic salvage logging and artificial replanting after natural disturbance.

- *Congressional protection of natural forests.* Prior to the ancient forest war (1984 ± 5 to 2014 ± 5), battles for forest conservation were primarily fought in Congress in the context of designating national parks and wilderness. The ancient forest war was fought mostly in the courts of public opinion and law. Congress was a minor arena, because it was politically divided, as (initially) was the public. In the future, Congress will legislate protection not only of big trees but of natural processes because the public will increasingly value them.

- *Challenging the fire-industrial complex.* The timber industry's last best chance to log public forests is in the name of "salvage" logging after a forest fire, a process named to even greater effect than ancient forests. The *fire-industrial complex* (an iron pentagon of the timber industry, government bureaucracy, old-guard foresters, elected politicians, and private firefighting corps) conspires for its mutual self-interest at the expense of both naturally functioning forests and the public treasury. Tighter fiscal times, increased scientific knowledge, and public outrage as this racket is exposed will bring dramatic changes to forest management.

- *Three decades and you're out.* At most the timber industry has three decades before any significant commercial logging ends on federal public lands. Social attitudes toward forests—especially public forests—will continue to change. The ever-urbanizing public will value scenic and ecological function more than wood products and stumps. The public will increasingly value the federal public forests for goods and services that the private sector is unable or unwilling to provide. Supply and demand affect the price of wood, but the social effects on price are even greater. As magnificent forests become more rare, they are more valued by the public for reasons other than board feet and chips. Economically, socially, and ecologically, timber production is no longer the highest and best use of the public forests, where much of our remaining old-growth forests reside. Such will increasingly be the case also with private timberlands.

- *Three decades of public forest restoration.* Big Timber's last three decades will trace an inevitable decline, but this final decline will be deferred because of the ecological necessity of often (but not always) using chain saws to repair and/or remove unnatural configurations and amounts of wood from the forests. Monoculture plantations in wet Westside forests can often benefit from variable-density thinning to accelerate the onset of late-successional characteristics. Similarly, many dry Eastside forests have unnatural buildups of understory trees due to high-grade logging, fire suppression, and livestock grazing. Reintroducing fire to these ecosystems is critical, but the danger from otherwise beneficial ground fire being carried into the residual old-growth canopy via this unnatural buildup of ladder fuels is significant. Thinning from below is often ecologically desirable. Industry can profitably thin both plantations and fire-suppressed stands on federal forests, and this should be done sooner rather than later. However, the backlog of excess biomass won't last forever, as forests will be regulated more by nature than chain saws.
- *Trees grow more slowly than money.* Increasingly, fiber is fungible and products traditionally made of wood — such as construction and paper products — have substitutes, some often bad (steel, petroleum-based plastics, etc.), and some good (agricultural wastes now mostly burned, etc.). The unsolvable problem of forestry is the time-value of money. As the timber industry moves from its hunter–gatherer phase (find a forest and clearcut) to its agricultural phase (plant fiber, wait, and harvest), it finds shorter rotation fiber crops to provide better returns on investment than long-rotation trees. Fiber production will move to the farm from the forest.
- *Marketizing nontimber forest values.* Economic markets will develop or be refined to value biodiversity, water (both quality and quantity), carbon, open space, recreation, habitat, medicinal molecules, and other forest products. Market forces will apply more — but not exclusively — on private timberlands than public forest lands. An interesting economic and social question will be whether it is more efficient for the public to obtain nontimber forest benefits by subsidizing and otherwise encouraging such markets or by simply reconverting private timberland to public forest land, thereby removing these areas from the pressures of the market.

Conclusion

The days of logging old growth are effectively over, for it has become politically incorrect to log old or big trees. The debate on the future of old growth has waned, although there is still some fizzing and fuming. The chore now is simply to mop up the pockets of resistance. Although conservationists remain ever watchful of the primeval forests they have helped remove from the path of the chain saw, other battles and other bars await.

LITERATURE CITED

Franklin, J. F., K. J. Cromack, Jr., W. Denison, A. McKee, C. Maser, J. Sedell, F. Swanson, and G. Juday. 1981. *Ecological characteristics of old-growth Douglas–fir Forests*. General Technical Report PNW–118. Portland, OR: USDA Forest Service, Pacific Northwest Forest and Range Experiment Station.
Seideman, D. 1990. Terrorist in a white collar. *Time* 135(26):60.

Chapter 12

A Private-Lands and State-Lands Perspective

HOWARD SOHN

Task number one for anyone offering commentary on a topic like "old-growth" forests is to acknowledge vantage point—to disclose relevant personal history, experience, and stake in the subject.

I grew up the son of a sawmill owner in the 1950s in Roseburg, Oregon—a rural community that, with some hyperbole, dubbed itself the "timber capital of the nation." Logging and mills dominated Roseburg's economy and provided a large share of the nation's growing demand for wood building materials.

My father's conservation ethic expressed itself in manufacturing innovations that resulted in more product from each tree and big reductions in waste. And he applied the latest findings in forest science to converting logged-over lands to productive, high-yield forests.

After high school, I left Oregon. I returned to Roseburg thirty years later (following stints in academia and corporate management) to help manage our family's wood manufacturing and timber-growing company. In 1990, the year of my return, the northern spotted owl was listed as a threatened species under the Endangered Species Act. The debate about forest use versus forest protection was in full swing. The spotted owl was the icon of the moment, a surrogate for old-growth forests, considered by some to be on the verge of extinction themselves.

Most of the old timber on private lands was gone, replaced by younger forests. On federal lands, over thirty years of commercial harvest had reduced acreage in the old-growth age class. Somewhat less than half of our million-acre Umpqua National Forest had experienced an initial harvest, and that portion of the landscape was now in reforested stands of varying ages. Substantial old growth remained, but in a more fragmented configuration considered detrimental to several wildlife species but beneficial to others. Roads that provided access for timber harvest and recreation were now plentiful and, because of their intrusion on hydrologic processes and sediment delivery to streams, potentially detrimental to aquatic habitat.

Harvest practices rules in Oregon had gone through several iterations since the Oregon Forest Practices Act of the early 1970s. Although older road-building specifications and harvest practices had been replaced by more environmentally sensitive methods, and reforms in these areas were continuing, the legacy of earlier standards remained.

In short, my return to Oregon to manage the family forest products business coincided with a major shift in forest management policy. By 1993, the spotted owl and old-growth ecosystem debates dominated policymaking, resulting in President Bill Clinton's Northwest Forest Plan, which redefined the purpose of the area's federal lands—from a not-very-well-balanced, multiple-use principle to a not-very-well-balanced, old-growth reserve and restoration strategy. Other important developments during the 1990s included reform of the Oregon Forest Practices Act and creation of the Oregon Plan for Salmon and Watersheds. Policies on management of state-owned lands also evolved during this decade within the context of the old-growth debate.

During the second half of the 1990s and into the early years of the new millennium, I became active in forest policy deliberations, although principally in connection with private and state forest lands. At the same time, I was overseeing a business in transition from a manufacturer heavily dependent on logs from federal forests to a private timberland manager and log producer without manufacturing facilities. My comments will therefore focus principally on the private-lands and state-lands perspectives on old growth.

An Industrial, Private-Lands Perspective

Since federal adoption of the Northwest Forest Plan and the resulting reduction in total commercial timber output in Oregon to about half of its peak levels (from 8 billion to 4 billion board feet per year), private forests have become the area's primary source of raw material for the wood products industry. Private forest harvests have increased only moderately, but their role has changed from a big player in the supply arena to the main player. Smaller, nonindustrial owners are important also, but not on the same scale.

Manufacturers accustomed to the larger supply from public sources faced a new situation. Logs for their operations would have to come from thirteen million acres rather than twenty-eight million. If federal land harvest restrictions persisted, they would have to achieve higher volumes of output from those fewer private acres. That meant enhancing productivity through management tools such as improved seedlings, fertilization, optimal density and, equally important, growing each crop only as long as its annual volume increment remained high—in other words, conversion to an agricultural model.

This was not a new concept; many private land managers were already following this course. But the urgency was magnified by the limited supply. The logs available from such a management regime would be smaller, because once trees in most western Oregon forests reach about age fifty, their growth rate slows and annual productivity per acre drops in comparison to what a new young forest on the same site would yield.

Shorter rotations, more uniformity, more aggressive management—all of this had ecological implications. It meant that private commercial forests would still provide lots of certain kinds of wildlife habitat but less of other kinds—lots of post-harvest clearings (good news for many birds), lots of young forests with fairly early canopy closure, but fewer landscapes with characteristics of mature forests. Forest practices rules, strengthened in recent decades to protect stream quality; reduce adverse impact from roads; and leave more snags, standing trees, and down wood after harvest, are all intended to ensure that these lands continue to contribute to forest habitat. But the rules would not provide the same habitat as what older forests contributed, in kind or degree.

This seemed acceptable because somebody else was providing older forest habitat, reducing the pressure for private lands to provide it. We each have our roles. The feds were specializing in older forests and habitat for species that need them; private forest owners would specialize in younger forests and species that prefer them (including humans who use wood).

So how do most private timberland owners look at old growth? Nice to have, important to have, but not our job. There were, of course, conservationists who argued that the new federal standards should apply more broadly. Industrial land managers countered that federal managers could learn from private managers who grew new forests in less time by applying the best human know-how and management tools.

The important-to-have-but-not-our-job view is not the whole picture. Even before Oregon and the Northwest saw the federal forests largely exit from the wood supply business, the region had lost its market dominance and with it the competitive advantage that accompanies dominance. Other advantages remained, including tree species and timber quality.

That situation was further aggravated by the dedication of federal forests to old age classes and the resulting supply diminution. Would the region still have enough logs to fuel a competitive industry? Would technologically advanced manufacturing, with its increased need for high volume throughput, have enough supply at its disposal to justify the investments necessary to stay competitive? Would an industry relying on less than half the productive acreage be large enough to support the necessary infrastructure—suppliers, transportation, knowledge base, expertise, skilled work force? Would communities hit hard by mill closings and reduced employment at good wages find other ways to maintain economic health?

Communities dependent on the forestry economy did take a severe blow, and the impact went beyond reduced employment. To the extent that communities were able to attract new businesses, they were primarily in sectors with lower wages. Further aggravating the surviving wood products businesses—and their ability to attract employees—was the resulting decline in public confidence that wood products careers were sustainable.

So it really did matter, this policy that said federal forests in the Northwest are in the old-growth business, the old-growth-dependent species business, and the forest conservation/preservation business. Protection of old growth had become a statement about what federal forests were for. It meant that, going forward, they would not be a significant part of the commercial supply picture. Even the degree to which they wanted to be players in the commercial arena (via thinning in late-successional reserves, harvesting in the matrix, salvaging after fires) was largely precluded by the "old-growth" advocates and ideologues who went to court to stop the limited harvest provided for in the federal management plans.

So advocates who fought under the old-growth banner became (or remained) the enemies of a private sector that did not reject the need for older forests but believed old growth was only one feature of a more diverse forest landscape that must serve society's multiple needs.

Although many private land managers subsequently moved to the sidelines of the debate, saying we don't have a dog in this fight, others looked at the bigger picture and said it still matters. We have a short-term scarcity, and we are coping with it now, but long term we all lose together if we aren't sizable players in the market. And we can't be big players if more than half of the forested acres (fifty-seven percent in Oregon) are largely off the supply table. Many took the position that the federal land managers were irresponsible in their failure to use the full range of forest science and forest management tools available to them. If they had, federal forests would remain part of the supply picture at some level.

An Oregon State-Lands Perspective

Management of state lands in Oregon illustrates what is at issue and how various perspectives and interests interact. The almost 800,000 state-owned acres comprise about three percent of Oregon's forest lands overall and about five percent of its public lands—not a huge factor, but because of its geographic concentration, very significant to some regions, ecologically and commercially.

The biggest ownership, the Tillamook State Forest in Oregon's northwest corner, was devastated by fires in the 1930s and 1940s. A major reforestation effort in the 1950s and 1960s resulted in a maturing forest by the 1990s. Its maturation, after decades of growth and little harvest, triggered a strategic management review by the Oregon Board of Forestry, the citizen body charged with management of these lands in accord with relevant state statutes. I served on that board, one of seven governor-appointed members (three of whom may earn their living from forest-related work), for eight years. During this time, a central task was development of a management plan for this state ownership. Although the board members come from particular places and sometimes particular interests, their role is to shape a plan to meet the broad array of public interests in optimal fashion.

That is a tall order. Deliberations are carried out in public meetings that include public input from the full range of interests. Some wanted to place much of the acreage in preserves, precluding harvest (other than temporary thinning programs), so that the forest might become "old growth." At the other end of the spectrum, commercial interests sought optimal timber production to generate employment and community revenues. The counties had originally deeded these lands to the state with the expectation of future timber revenue to fund county services.

The board also heard from moderate voices with more nuanced positions, as well as from scientists and other experts from across the spectrum.

It undertook and commissioned studies to assess the forest's capacity to meet an array of desired outputs, including environmental, commercial, public revenue, recreation, and aesthetic values. The board's solutions also had to comply with its constitutional and legislative mandates for these lands—mandates originating in another era when prevailing values and perspectives on natural resources were not of the same mix as today.

The Oregon Department of Forestry, along with the board, developed a plan that sought to optimize the forest's capacity, within the board's mandate, to address the range of values and desired forest outputs—to meet, as they expressed it, an appropriate balance of social, environmental, and economic goals. Among the implicit, sometimes explicit, principles are the following: Symbolism is not important, outputs are; ideologies cannot be satisfied, but legitimate public values can be addressed; current practices and outputs are important, but long-term results—a century or two out—are primary. A few examples include that: No single icon, such as the northern spotted owl or "old growth," will drive planning; value resides not in the age of trees but in habitat characteristics; productivity is not measured only in timber volumes or dollars but also in recreation units and habitat features; streams are not healthy because (or if) they are untouched but because they exhibit the physical and biological characteristics that support fish; roads are not bad per se, but adverse hydrologic impacts are.

The management plan that emerged fell about midway between the protection-dominant scenario now in place on federal lands and the production-dominant regime common on private industrial lands. Not that it just split the difference in compromise. On the contrary, it found a "new way" that said here is a way to manage the whole landscape that contributes substantially to biodiversity of the region's forests, including many (not all) old-growth forest characteristics.

The plan had these assumptions: (i) The value of old growth is in its habitat characteristics—its capacity to support certain old-growth–dependent species— and (ii) achieving those results will be pursued actively rather than passively, using tools developed by forest scientists. On these assumptions, the plan set out to manage this young forest so that over time portions of it would contribute old-growth habitat characteristics without sacrificing other forest values as permanent set-asides would. A further assumption was that an 80–120-year-old forest, managed to this purpose, could achieve substantial old-growth character. If the state is to meet the full range of objectives over this relatively small forested acreage, the best way would be to rotate most of the acreage (with the exception of special areas) through the various stages of the natural cycle. With respect to old-growth habitat characteristics, this means that as some portions of the forest mature to the point where such characteristics are prevalent, older portions of the forest where these characteristics have been present for some decades might then be converted to young forests through harvest, and the cycle would start again.

Are these older, longer rotation portions of the forest true old-growth forests? Not if the definition of old growth is a stand of at least 180–220 years of age. Not if the definition is a forest with the complexity of a 400-year-old forest that has experienced only low-intensity fires. But if the distinctions between old-growth and non-old-growth are less absolute, and if we want old-growth forests for their characteristics—large trees, various species, multiple canopy layers, down wood and debris, snags and broken tops, streams with complex structure—then these forests qualify as contributors. They are a substitute; they do not provide what much longer rotations (or no managed rotations) provide. Instead, they are a valuable complement, providing a significant portion of the suite of old-growth values.

There was and is substantial opposition to this management plan, not only from those who believe this contribution to old-growth habitat is insufficient, but also from those who believe that these state forests have other purposes that will not be achieved with long rotations. The plan will not maximize revenue and will produce—from 80–120-year-old stands—logs larger than most modern mills can process.

The Board of Forestry believed, however, that state forests have a role that differs not only from federal forests (to which one set of critics often compares it) but also from private industrial lands that focus on log output (and to which another set of critics often compares it). Everybody wants the state forests to serve their interests first and most. However, they cannot, given their size and mandate, be true old-growth forests, but they can contribute old-growth features to the forest landscape. They cannot, given their size and mandate, provide industrial forest harvest levels, but they can contribute substantial commercial volume from across their entire landscape over time.

Some Derived Principles

What principles, relative to the old-growth debate, emerge from these two perspectives—that of a private industrial landowner and that of the managers of Oregon's state forests?

Perhaps the most prominent principle, and common to both perspectives, is that an active human role in managing the landscape is preferable to a passive approach. Letting nature take its course is not a practical or responsible option once the landscape is inhabited to the extent that it is today and will be in the future as residential encroachment continues. The knowledge and tools developed by forest science should be applied in the pursuit of desired forest outputs and values. Not only is the allocation of the landscape among differing objectives and uses a human choice, but the optimal achievement of all these objectives depends on the application of human knowledge. This is not to deny that knowledge and tools will advance over time, but waiting for some final state of knowledge is not a realistic option.

Another principle is that human use and species habitat are not mutually exclusive. Although part of the answer to allocating uses may be to dedicate portions of the forest to one primary purpose at the expense (but not necessarily exclusion) of the other, much of the forest landscape can be managed for multiple objectives.

A principle that has more to do with process is that nobody's rights are absolute. Forest property owners must consider the public impact of their land management, and environmental advocacy professionals, who often claim to represent the public interest, must acknowledge the ownership rights and legitimate commercial interests of landowners.

Contrary to the impression often created in the old-growth debate, the public interest and private interests are not necessarily at odds. The public is not synonymous with those who claim to be its advocates. The public benefits from the forest in a wide range of ways, from availability of wood products to recreational access to old forests largely precluded from commercial use. As more of the forest is allocated to noncommercial and residential uses, the remaining portion will carry an increasingly heavy burden of meeting society's needs.

Another principle is that old growth for old growth's sake is not a value; the value is in its ecological characteristics. Nor do the private- and state-land perspectives deny that the values of old growth may exceed or differ from what we know them to be at this particular—or any particular—time. But it remains incumbent on us (and on forest science) to ascertain more fully what and how old growth contributes and then foster those contributions. That probably doesn't mean protect and foster to the maximum but rather to a degree proportionate to the value of that contribution relative to other values.

In other words, the-more-the-better principle does not apply here. To turn briefly to the overused concept of *sustainability*, a balance of use and protection (not use *or* protection) is sustainable, whereas an imbalance is not. Sustainability ultimately rides on political (public) consensus. An old-growth forest in reserve status is sustainable only if enough of the forest resource is available to meet other needs. An optimal level of older forest characteristics in nonreserve, actively managed forests is sustainable only if in proper proportion to other demands.

Reflections

For the most part, discourse over old growth and the policy issues it stands for has been neither a debate—in the sense of a presentation of arguments objectively judged for their strength—nor a dialogue—a mutually respect-ful exchange of views. In public forums and the press, it has been a rhe-torical contest, and in the courts, a battle of interests waged with techni-calities of the law as weapons. At least that is how it has looked from my vantage point. The challenge is to create academic, policy, and community forums in which intellectually objective and mutually respectful dialogue and debate take place.

What I have seen is largely posturing for advantage, a pursuit of inter-ests driven by ideology or by stake, by how the participants' interests are affected. It is rare that a participant acknowledges elements of validity in an alternative viewpoint. It is unusual that references to research support for a position also acknowledge the existence of research with conflicting con-clusions. It is unusual to find participants who are not aligned with a "side," or at least perceived as aligned with or beholden to an interest.

It is not surprising that participants in the debate will be those with a stake in it. Disinterested parties are rarely engaged or informed. The excep-tion may be academics, who bring a level of objectivity, along with expertise beyond that possessed by advocates. Even academics are often perceived to be aligned (or are aligned) with interest groups. Although the hope for greater objectivity rests with the scientists and academics, they can only inform the debate about policy. Policy directions and decisions are driven largely by values and priorities, not by science.

Finally, a comment on the role of symbolism in the debate and the vocabulary that has developed around it. So much of the discourse is addressed to and presented by the media, which delivers in headlines and shorthand. The media need images—verbal and visual images— that carry a lot of meaning in a small space. The terms *old growth*, *ancient forest*, *virgin forest*, *pristine forest*, *untouched forest*, *roadless forest*, *natural forest*, and *wilderness* all carry big messages in a small space. The problem is that they oversimplify and appeal more to emotions than intellect. They rarely serve informed, objective discourse.

As the old-growth advocacy message has morphed (and the definition of old growth has evolved to include younger and younger trees) into a message about trees and forests in general, the call for saving old growth has become a call to save "trees." Use plastic, save a tree; make two-sided copies, save a tree; recycle, save a tree. Using paper and wood products that come from trees has entered the culture as guilt-worthy activity. Conservation is essential, but attaching the emotional value—even the environmental value—of old growth to trees in general is an unfortunate effect of the old-growth debate. Instead, use of renewable resources—conserving, not wasteful, use—should be championed as part of a sustainable resource policy and lifestyle.

Chapter 13

Getting from "No" to "Yes": A Conservationist's Perspective

RICK BROWN

Our current political climate seems locked into a self-reinforcing loop of polarization—both feeding on and fostering sharply divided, entrenched positions. Debate over protection of old-growth forests has not been immune from this tendency. The stark visual contrast of awe-inspiring stands of old-growth Douglas-fir juxtaposed with harsh clearcuts gave conservationists an effective advocacy tool, but it also may have contributed to dichotomous either-or thinking about old growth: Either fully protect it, or it will be gone.

Humans, Old-Growth Forests, and Change

It used to be simple. Never easy, mind you, was this task of protecting old-growth forests, but straightforward—define it, map it, draw a line around it, and leave it alone.

That's an exaggeration, of course, but maybe not a big one. In the 1980s, as scientific evidence confirmed that there was something special about the ancient, cathedral groves of western Oregon, Washington, and northwestern California, the overwhelming threat was immediate loss through clearcut logging. If we could stem that destructive tide, then there

149

would be plenty of time to sort out the niceties of how to hedge against losses due to wildfire or other natural events and how to replace old growth over time. As with so many environmental issues, delay by resistance was a key strategy. Executed well, it could allow political reality to catch up with science and public opinion. And the fact is, if only they are given a chance, old-growth forests will wait while we humans sort out our differences.

The dominant trees in the famous forests of western Oregon are so long-lived that they are effectively irreplaceable. The forests, like all living systems, are as much a sum of the ebb and flow of processes as they are objects whose dimensions can be defined. However, the overall rates of change in the most evident life forms are sufficiently slow that the general appearance of a given stand can seem unchanged over the course of decades. Events that might dramatically alter its appearance, such as a severe fire or extreme wind storm, typically occur in forests on the west side of the Cascades only at intervals exceeding several human life spans. Even the more common form of change, the falling of individual large trees, entails subtle alterations that leave both ecological functions and the overall impression intact.

Human resistance to change is the basis of much environmental action and most of the resulting contention. Conservationists want the places, species, and evolutionary possibilities they cherish to persist, while commodity interests such as the timber industry (and the communities that rely on it) want a steady and predictable supply of logs. Both sides may acknowledge that change is inevitable, but each would like to pace the rate of change to suit its own preferences.

Although protection of timber was the primary purpose of the fire suppression program, conservationists long supported the program in all the nation's forests, seeing uncontrolled wildfire as a threat to their favorite hiking trails, water quality, or critters such as the northern spotted owl that depend on old, green, growing forests for their survival. Now, as many have come to accept that fire can have an appropriate role in the forest, others seem to see fire as a benign solution to all the forests' problems: "Burn, baby, burn" has emerged as the new mantra in some circles. Change, it turns out, can become a very slippery thing to embrace.

Turning Westside Eyes to Eastside Forests

As battles over forests within the range of the spotted owl slowed down in the mid-1990s with the adoption of the Northwest Forest Plan, some Westside conservationists turned their attention to the conservation of old-growth forests east of the Cascade Mountains, particularly the dry, low-elevation forests historically dominated by ponderosa pine. They were, of course, joining others who had worked in and for the Eastside forests for many years, quietly toiling in the shadow of the publicity around spotted owls and Westside old growth.

Many of the issues in the eastern dry-side forests were familiar: huge, long-lived, and aesthetically appealing trees; habitat for distinctive wildlife; and disconnected remnants representing a small fraction of the forests' historic distribution. Despite these similarities, however, there were also some important, if not always obvious, differences.

Two defining elements of Westside old growth are that they have multiple canopy layers and several species of trees. Vast areas of the Eastside (and the rest of the interior West) consist of open forests of ponderosa pine—one species of tree, one canopy layer, and (seemingly) one age class. Typically, fire did not disrupt or disturb these forests but rather maintained their old-growth character by frequently clearing seedling trees, shrubs, and accumulated debris from the forest floor, as well as pruning the pines' lower branches. Predominantly single-storied old-growth expanses are interspersed with patches of younger trees, yielding a less-concentrated complexity than Westside old growth but one just as rich when considered on a landscape scale. Where conditions are more moist, such as higher elevations or north-facing slopes, characteristics of forests and fire regimes are more complex, but ponderosa pine old growth is the archetypal Eastside forest.

Although some such stands still persist, the ponderosa pine landscape has been transformed by decades of logging the big trees and excluding fire, first by the removal of fine fuels by livestock grazing, then by active fire suppression. The absence of fire allowed many more young trees to establish and grow. Even in areas where old ponderosa pines still remain, they often struggle to coexist with unprecedented understories of young pines and other conifers such as Douglas-fir or true firs. The understory trees compete with the overstory pine for water and nutrients, stressing the old trees and making them more vulnerable to attacks by bark beetles. These smaller trees also serve as "ladder fuels," helping to carry fire into the crowns of the old pines, with lethal effect on trees that may have survived a dozen or more "natural" fires over prior centuries. In some respects, these forests have come to resemble classic Westside old growth.

At best, replacing these old-growth trees will take centuries, and climate change and associated adaptations in fire regimes may preclude reestablishing forests at all on some sites. With continued fire exclusion, the risks to such stands only grow worse, while uncontrolled fire is apt to wipe them out. Unlike on the Westside, extreme fire weather can occur east of the Cascades almost annually.

On the Westside, the urge to protect old growth by leaving it alone was compatible with the pace of ecological dynamics; on the Eastside, in drier forests, active intervention with fire, understory thinning, or both would be necessary before natural, historic dynamics could safely be restored. Restoration became the new protection.

Ironically, many of these highly altered ("unnatural") stands ended up as "designated old growth" in Eastside forest plans, with any cutting of trees discouraged, and the U.S. Forest Service little inclined to risk a prescribed fire that might too easily burn out of control. The distinction between logging *of* old-growth forests, as opposed to logging *in* old growth to remove young ("unnatural") trees, was perhaps too subtle to be captured in management guidance as broad as that in a forest plan. Or maybe public support, even for thinning small trees among the old growth, had become too elusive, the trust eroded by too many years of perceived betrayal.

Challenges to Eastside Conservation Strategies

Adapting the conservation strategies used to protect Westside old growth for Eastside forests has not been easy for environmental activists, in part due to the very success they had in pursuing those strategies.

Simply coming to a full realization of the different dynamics (as well as structure and composition) of Eastside old-growth forests was an essential first step. Those with "Westside eyes" tended to look at fire-excluded, heavily ingrown dry forests and see the multilayered, multispecies stand as classic and desirable (Westside) old growth. Their reluctance to agree to any logging in old-growth forests was reinforced by those instances in which projects carried out in the guise of restoration logged old, fire-resistant trees in addition to the smaller trees that contribute to risks from competition and uncharacteristically severe fire. A distrust of politicians and the agencies they controlled, founded on decades of destruction of old growth, was understandable, if not always productive.

A second impediment to restoration thinning has been a desire to rely on "natural" solutions—in this case, fire. Some have assumed that if the forests we seek to restore were historically created and maintained by fire, then allowing fire to burn "naturally" would be the best way to restore them. This focus on "naturalness" fits in well with a distrust of the Forest Service, perceived as an advocate of commercial logging and an aggressive squelcher of fires. Given enough time—many hundreds of years, at a minimum—fire and old-growth forests likely *will* find a new equilibrium, but existing old growth would be long gone before replacements could develop and mature. Fire is certainly part of the answer, but it is not the sole answer.

A third trend working against Eastside restoration was the increasing size and severity of wildfires feeding on fuel that accumulated while fire continued to be excluded from the forest. These fires posed a threat not only to fuel-enriched stands of old-growth ponderosa but also to homes and communities that were spreading into forestland adjacent to national forests. Political attention to these problems was perceived to run a great risk of leading to unlimited logging in the name of reducing the threat of fire to homes. Deciding that it wasn't feasible to just say "no" to fuels reduction, but fearing widespread logging, many conservationists decided to become staunch advocates for fuels reduction treatments but only in areas adjacent to communities—the *wildland–urban interface*. Not trusting the Forest Service in the forests, the defenders of trees paradoxically became the defenders of houses.

Those who worked closely on conservation of these dry forests came to appreciate some additional differences—social differences—from the Westside. There, the timber industry and its supporting communities could be portrayed as the opponents of conservation, trying to continue their inroads on now sacred old-growth stands. On the Eastside, however, where thinning was a necessary element of the conservation solution, some form of timber industry was essential. Communities that could supply the expertise to work in the woods and to make products from the material removed would also be an integral part of the conservation picture. In too many places, there's just too much wood in the woods; it needs to come out, and if it can be used to create durable, value-added products rather than burning it and sending more CO_2 into the atmosphere, so much the better. For the near future, these forests need tending, and society needs to supply the expertise, infrastructure, and economic incentives to ensure that the tending happens.

A key point needs to be emphasized here, one with which many environmental activists themselves struggle: The fact that some forests need our thoughtful, caring management should not be construed as support for those who suggest that all ecosystems everywhere somehow depend on human intervention. This tension explains some of our hesitation in regarding as partners some of our previous adversaries.

Getting to "Yes"

In theory, it would be desirable and equitable to have the American public pay for the investments now needed in these forests, given that they bene-fited from low prices for logs from public land over many past decades. But the realities of current federal budgeting—war, homeland security, deficits, and national debt to the horizon and beyond—will put a premium on get-ting commercial byproducts from restoration thinning.

Over the past five to ten years, here and there around the West, advo-cates for community-based forestry, community representatives, commer-cial interests, conservationists, and agency personnel have been coming together to explore new relationships with each other and with surround-ing forest landscapes. It's been a cautious and tentative dance in most cases, gradually building a common understanding of the needs of forests and people as well as a vision of how to meet those needs. Funding, scientific information, markets, and appropriate equipment have all been in short supply, but often the ingredient that has seemed most limiting has been conservationists willing to risk building trusting relationships with old adversaries and in particular to advocate—against all their best instincts—for cutting trees down and taking them out of the forest.

For decades, forest activists have largely measured success by the num-ber of things they were able to successfully say "no" to—roads, stumps, pesticides, ski developments, mines, you name it. Now, we have to learn to add a "yes" to our vocabulary occasionally, most notably the "yes" involv-ing cutting trees in the interest of restoration. We're not finding it easy. One friend, whose organization was taking some very tentative steps toward supporting small restoration thinning projects, exclaimed, "Man, I liked it so much better when we could just say 'no!'"

Progress is being made, however. Stories of successful collaborations are coming in from around the West, and networks help exchange knowl-edge of what's worked and what hasn't.

My own understanding, however imperfect, of how to think about these forests and what needs to be done for and with them began in the mid-1970s. Around the time I was laying out plots and transects, most of them in old growth, for graduate research in the western Cascades, I had my first opportunity to visit Eastside forests with a knowledgeable and thoughtful ecologist who helped me perceive the changes wrought by fire exclusion. A subsequent, unexpected stint as a biologist on the Mount Hood National Forest provided opportunities to spend time in the ponderosa pine forests along its eastern edge. Then, working as staff for the National Wildlife Federation, the "spotted owl wars" kept my focus firmly on the Westside from the mid-1980s to early 1990s.

But the lure of the dry side was not to be resisted. Forest health and other initiatives from the Forest Service provided opportunities, but they tended to deal with the abstractions of policy rather than the particulars of a given place. In 1998, I went to a meeting in the town of Lakeview in south central Oregon, primarily to say "no" to the idea of green certification of national forest lands in the vicinity. Conservationists questioned the appropriateness of applying certification standards—originally developed to recognize better-than-usual management on private commercial timberland—to public lands that have very different goals and objectives. We were also concerned about the certification process's lack of a system of public accountability comparable to what otherwise exists for federal public lands. Although the plan for certification went away (for the time being), I was introduced to some community leaders who realized that times had changed, that restoration with timber and other community benefits as byproducts was now the only way to go, and that conservationists would be necessary (if perhaps unsavory) partners.

Thus began my involvement in and commitment to a collaborative, restorative approach to a piece of fire-prone forest on the Eastside. Our efforts have focused on what began as the Lakeview Federal Sustained Yield Unit, approximately 600,000 acres of the Fremont National Forest where local mills get, in effect, a right of first refusal on any timber sales (and timber sales were the name of the game here for several decades). Much of the landscape in what has been renamed the Lakeview Stewardship Unit consists of dry forests historically dominated by ponderosa pine.

Participants in the collaboration (the Lakeview Stewardship Group) include conservationists primarily from the Interstate 5 corridor cities of Seattle, Portland, Eugene, and Ashland; representatives of the one remaining local mill (owned by The Collins Companies, widely recognized for exemplary management of their private forest lands); and local civic leaders and citizens. Conservation of old-growth forests is implicit in our broad goals addressing habitat, restoration of fire, and ecosystem resilience. Forest Service staff are active participants and important sources of information. The stewardship group has recently completed a long-range strategy for management of the unit, which we hope will both provide guidance for project-level strategies and inform the process of revising the Fremont's forest plan, due to start next year.

Adjusting Conservation through Collaboration

It's rarely easy and it's never simple, but collaboration seems to hold more promise for enduring solutions than anything else I've come across. We don't have all the answers—far from it—but we have some, enough to move ahead while we try to figure out some more. We're trying to hold onto the old-growth ponderosa pine still here and provide for more in the future. The greatest challenges are not with the always humbling ecology of these systems but with the sociology and economics of how to get restoration done. How do we gain support in a busy, cynical, and distrustful world? How can we figure out the range of uses, from sawtimber to biomass, to optimize the economic benefits of ecologically driven restoration? The puzzle is gradually being filled in, in part by sharing experience and knowledge with people involved in similar efforts around the West.

Recently, a conservationist with a long history in the Northwest "forest wars," widely and justly considered to be visionary, published an online essay declaring that forest restoration and collaboration with the Forest Service are now the way to go. Although on one hand I'm tempted to express hope that he can catch a train that's long since left the station, I'm also more than happy to welcome him aboard. There are still a lot of empty seats.

Then again, a couple of days after I read that piece, I saw an article about collaborative efforts in northeast Washington and Montana. Here conservationists were quoted saying that they were unwilling to discuss thinning more than a quarter mile from houses or in old-growth stands, because this would be "too controversial." For some, controversy biology still trumps conservation biology, but at least we've made a start. This time, however, I'm not sure the forests can wait for us to get our act together.

Conclusion

Our resistance to change itself needs to change. Although it's a canard that conventional conservation strategies are premised on views of nature as static, it can sometimes be a challenge for conservationists to accept just how dynamic ecological systems can be. Even those who recognize this dynamism may be reluctant to acknowledge it out of concern for how this information can be abused by those who will suggest that natural change gives license to *any* form of human-induced change. We need to be clear that some changes are acceptable, some are not. As our society haltingly comes to terms with that other mind-bending change we face—global climate disruption—we can only hope that our ability to accept and plan for change will start to catch up with our ability to cause it.

Chapter 14

Old Growth: Failures of the Past and Hope for the Future

ROSS MICKEY

The famous John Godfrey Saxe poem about the six blind men and the elephant aptly applies to the "old-growth" debate:

> It was six men of Indostan
> To learning much inclined,
> Who went to see the Elephant
> (Though all of them were blind),
> That each by observation
> Might satisfy his mind.

The first blind man feels the elephant's side and proclaims that an elephant is like a wall. The second touches the tusk and cries that it is like a spear. The third takes hold of the trunk and says it is like a snake. The fourth feels the knee and says it is like a tree, and the fifth touches the ears and declares that it is like a fan. The sixth disagrees as he grabs the tail and says it is like a rope. The poem ends as follows:

> And so these men of Indostan
> Disputed loud and long,
> Each in his own opinion

Exceeding stiff and strong,
Though each was partly in the right,
And all were in the wrong!

(Saxe, 1892)

The old-growth debate is littered with analogous situations, chief among them that there is no agreement as to what old growth is, how much exists, or how much existed in the past, although many proclaim "loud and long" that they know. There is also no basis for determining if we have too little, just enough, or too much of it, whatever "it" is. Lastly, there is no clear understanding of the importance of this thing we call "old growth" in the ecological, economic, and social sense. Even though the entire subject of old growth is vague and ill defined, there is no lack of people adamantly proclaiming facts and positions as if they really know what an elephant is.

Old Growth—A Thousand Definitions

My literature search turned up scores of definitions of old growth—*scores* (Franklin et al., 1981, 1986; Jones, 1988; Marcot et al., 1991; Potter et al., 1992; USDA, 1992, 1993; White and Thomas, 1995). These are usually based on sets of ecological characteristics such as tree species, plant association, geographical region, soil productivity, elevation, site class, and tree age and origin. One would expect different definitions for different tree species, but even when you narrow down the discussion to one plant association such as Douglas-fir forests in the Pacific Northwest, there is still no agreement on how to define it.

In the most recent assessment of the status of old growth within the range of the northern spotted owl, the authors used three different definitions to estimate how much old growth might exist under each one (Moeur et al., 2005). Depending on which definition they chose, they found a difference in abundance, patch size, and distribution of "old-growth" Douglas-fir in the Pacific Northwest.

The three definitions and estimates of how much old growth might exist under each were

- Older forest with medium and large trees and single- or multistoried canopies—thirty-four percent of forest-capable area (7.87 million acres ± 1.96 million acres)

- Older forest with medium and large trees defined by potential natural vegetation zone—thirty percent of forest-capable area (7.04 million acres ± 1.93 million acres)
- Older forest with large trees and multistoried canopies—twelve percent of forest-capable area (2.72 million acres ± 0.35 million acres).

Who gets to decide what the proper definition is? It appears that everybody does. This is why we have different definitions and abundance estimates from the U.S. Forest Service, the Bureau of Land Management, several universities, the Wilderness Society, the Society of American Foresters, and many remote-sensing contractors. Which one is right? Maybe the "blind man" knows.

That there are so many definitions and that each definition is specific to a certain tree species, plant association, geographical region, etc., although interesting to scientists and researchers, tends to overwhelm the general public and further muddies the water of the public debate. Newspapers and television reporting aren't conducive to clarifying complex issues. That is left to the PBS special series aired late at night. What sells on the evening news are ten-second sound bites that grab peoples' emotions. What you may hear are things such as, "We must stop the Forest Service from logging the last of the ancient cathedral forests. Call your congressperson and tell him or her to defeat HR-XXX." It matters little to the local news reporter that no one has ever put a definition to "ancient cathedral forests," or that no one knows how much or where such forest may or may not exist, or that no one can say if the Forest Service is logging any or all of it. Despite this, the news report is aired, the general public is outraged, and donations pour into the group supporting the effort to "stop it."

But We Have Only Ten Percent of the Original Old Growth Left!

Some may say that this lack of a common definition for old growth is well and good, but no one can argue with the fact that we have only a small fraction of "it" left. This notion is held to be intuitively obvious because we have been harvesting (and growing) trees since settlers moved here. For some, this moment in history when settlers arrived is held as the standard for what we should have today or is used as the yardstick to determine how much we have lost. This, however, ignores many fundamental facts of ecological systems, one of which is that ecosystems are always in a state of change. The condition of the landscape when settlers arrived could have been at a low, mid- , or high point in the natural fluctuation of the system. If we don't know where in the cycle the landscape was when settlers arrived, how can this be used as a reference?

A few studies have tried to determine how the amount and distribution of older forests have changed over time. These have been done in well-defined geographical regions, using clearly stated definitions and specific points of historical reference. The majority of these are based on postsettlement records. The problem with using this approach is that it shows us only how things have changed over the past 150 years or so. This is a very short time frame when looking at long-lived species such as Douglas-fir that can grow for 250 plus years barring any catastrophic event such as windstorms and wildfires.

These studies give us no sense of how these landscapes naturally changed over the past 1,000 years or more. This longer time frame could represent about four natural cycles of a Douglas-fir forest. One study did look at a longer time frame using data from dendroecological and paleo-ecological studies to simulate temporal and spatial patterns of wildfires to model the historical variability of older forests in the Oregon Coast Range over the past 3,000 years (Wimberly et al., 2000). What the authors found, I believe, is applicable to most forest types and regions:

> Our results indicate that the historical age class distribution was highly variable and that variability increased with decreasing landscape size. Simulated old-growth percentages were generally between 25% and 75% at the province scale (2,250,000 ha) and never fell below 5%. In comparison, old-growth percentages varied from 0–100% at the late-successional reserve [see box 2.1] scale (40,000 ha).

What does this tell us? First and foremost is that over the 3,000 years modeled, all of the older forests were replaced at least once. Over the course of 3,000 years, every landscape smaller than 100,000 acres had either zero percent or 100 percent older forests, *naturally*. Only when you look at a very large area, more than 5.5 million acres, do you find that the low point goes from zero percent to five percent. So, it was only within these very large areas that older forests may have persisted, but only on a maximum of five percent of the area.

I think these data also show that it is not possible to use the past as a template for suggesting what we "need" today. If we tried to do this, what point in the fluctuation between zero and 100 percent would we choose as the baseline and why? This, incidentally, was the conclusion of Wimberly et al. (2000) for forest areas less than about 500,000 acres in size:

> Our results suggest that in areas where historical disturbance regimes were characterized by large, infrequent fires, management of forest age classes based on a range of historical variability may be feasible only at relatively large spatial scales. Comprehensive landscape management strategies will need to consider other factors besides the percentage of old forests on the landscape.

So, how much "old growth" remains? 10 percent? 50 percent? 75 percent? The answer depends on what point in time and at what scale you compare the past to the present. Is it reasonable to assume that there was a point in time when there was about the same amount of old growth as we have today? I believe the answer is a resounding "yes."

Be wary of the "blind man" proclaiming that we have only 1 percent, 5 percent, 10 percent of the original old growth left. You really don't know what point of the ecological timeframe he is touching or what definition he is using.

There Are Definitions and Estimates, Then There Is Reality

One thing that all estimates of the abundance and distribution of old growth have in common is that none of them actually uses the definition it claims to be using when it estimates how much old growth exists today. This is confusing, so let me give you an example.

One of the most recognized definitions of Douglas-fir old growth in the Pacific Northwest is found in PNW-447, a research note published by the U.S. Department of Agriculture Pacific Northwest Research Station (Franklin et al., 1986). This defines *old-growth Douglas-fir* on western hemlock sites as "stands" (undefined by the authors) having the following characteristics:

- Two or more species with wide range of ages and tree sizes
- Douglas-fir ≥ eight trees per acre of trees > thirty-two inches in diameter or > 200 years old
- Tolerant associates ≥ twelve per acre of trees > sixteen inches in diameter

- Deep, multilayered canopy
- Conifer snags ≥ four per acre that are > twenty inches in diameter and > fifteen feet tall
- Downed logs ≥ fifteen tons per acre, including four pieces per acre > twenty-four inches in diameter and ≥ fifty feet long

All of these indicators are easily measured if you actually go out into the forest. A group of people in the woods would have a pretty good chance of being able to agree if the spot they were standing in met this definition (disregarding the lack of a definition of a "stand.")

But what happens when we try to determine how much of this type of forest exists over a very large landscape? It doesn't take very long to realize that there is not enough money or is it physically feasible to have all of the "stands" in that area visited by real people. So how are estimates made of the amount of PNW-447 Douglas-fir old growth? The simple answer is computer simulations, models, and statistical estimations. Most commonly, spectral images taken from space are analyzed and computer algorithms are written to estimate forest characteristics that are surrogates for the real definition. Clear as mud, right?

Satellite imagery and computer simulations don't provide much detail. Technology beamed from space does not, for example, tell us anything about the size or age of the trees we are observing, key factors used in most definitions. Eyes in the sky can tell us what species they see, how big the crowns are, and a few other things, but you have to actually *walk* the ground to determine whether or not all six characteristics of PNW-447 old growth exist there.

So how much PNW-447 Douglas-fir old growth do we have in the Pacific Northwest today—and how much did we have before white people began their western migration? Well, we really don't know, but there are people out there who *think* they know the answer and proclaim it "loud and long."

Just Save It

Now that 21 million acres of federal land in the range of the northern spotted owl has been set aside, how are we doing in protecting the old growth we have today? What was the number one cause for the loss of older forests in the past decade in the Pacific Northwest? Logging? No. Urbanization? No. The number one cause of habitat loss for the spotted owl and older forests was catastrophic wildfire. The Biscuit Fire in southern Oregon alone consumed 500,000 acres of forest land, mostly in wilderness areas and "late-successional reserves" created to "protect" these areas. Large areas within this burn were classified as "old growth." The Forest Service estimated that 68,000 acres of spotted owl nesting and roosting habitat and 51,000 acres of spotted owl dispersal habitat were lost in the fire as well as 20,341 acres of critical marbled murrelet habitat and 131,604 acres of critical spotted owl habitat (USDA, 2004).

Why did this fire consume so much acreage? The major factor was the unnatural buildup of flammable material, that is, too many trees and too much brush. Some of the reasons that this situation was allowed to develop are that active management is prohibited in wilderness areas, and there was a lack of funding to pay for the needed fuels reduction work. In recent years, this problem was compounded by the fact that little or no management occurred in the late-successional reserves created to protect the forest that ended up in ashes. The result of these policy decisions to "save" the area was that it burned down.

I Just Want to Make Sure That We Will Always Have It

What is the best way to ensure that there will be older forests in the future? The answer one hears the most is to set aside all of the existing old growth in reserves. Within the range of the northern spotted owl, the Forest Service and Bureau of Land Management developed a management plan—the Northwest Forest Plan—that set aside huge blocks called *late-successional reserves*. These large blocks are not 100 percent old growth, but the idea was that they would grow into such over time. Based on this management scenario, the agencies estimate that the amount of old growth would increase from 8 million acres to 10.7 million acres in fifty years. Although this is good news for those people who believe that we are on the deficit side of our old-growth needs, it is a shortsighted approach. As one researcher put it, "(F)orest preserves are not like strawberry preserves."

What this means is that dynamic forest ecosystems cannot be held static like strawberry preserves can. With strawberry preserves, you can be fairly certain that the contents will remain unchanged as long as the jar is not opened. Forest ecosystems are in a constant state of change. The forest you walk into today will be vastly different from what will be there in fifty to 200 years. What will be the likely outcome of the millions of acres of federal land now in reserves? It is an absolute certainty that, given enough time, these older forests will change by catastrophic wildfire, wind throw, insects and diseases, or other ailments related to old age.

People tend to forget that all old forests were formerly young forests, and all young forests ultimately started from seeds taking root in soil. The only way we can ensure that we will have old forests tomorrow is to continually start new ones today. The only way to have new forests is to have them start out as seeds or seedlings in bare soil. The reserve system of management relies on unpredictable events to create new forests. Thus, there is no way to predict how much area of older forests we will have in the future. If we really want the best chance of having older forests forever, we must manage our lands to achieve that goal and continually create new ones.

What would this look like? A healthy forest, one with trees that have high crown to height ratios and adequate space to grow, has a much greater likelihood of achieving old age than an unhealthy one. Because we have removed natural fire from the ecosystem, hundreds of thousands of acres of our federal lands are in an unnatural state, far too dense to maintain their vigor, waiting to explode in an uncharacteristically large catastrophic wildfire. Many scientists believe that these dense forests may never develop the characteristics described in PNW-447 (Old-Growth Definition Task Group, 1986). These lands must be managed if they are to develop the characteristics found in the old-growth forests today. They need to be mechanically thinned and the residue treated so that there are enough nutrients and water to support the remaining trees. Then, we need to continually create new forests.

This type of an approach, however, will meet great resistance by those who want to perpetuate the myth that somehow a tree that sprouted from a seed that fell naturally from a tree is superior to a tree that was grown from the same seed in a nursery and is then planted in the woods. There are those who want us to believe that nature should be left to take its course without human intervention. Very few people would take this course of action if it pertained to letting a natural disease take its course without a doctor's intervention. I, for one, will go to the doctor if I am sick, and I will do as I am instructed.

It is time to let the forest doctors, called *silviculturalists,* do their work. Tell them what forest condition you want on what proportion of the landscape, and they will tell you how best to achieve it. The current mode of locking forests up and walking away is doomed to failure.

Now is the time for action. We cannot wait for the scientists to agree on a definition or for the federal government to allocate billions of dollars so we can finally do a *real* survey to see how much we really have. Even if we did, we have no yardstick to use to judge where we are on the continuum of "the historic range of variability." Neither can we look to "old-growth" species to give us the answers. The spotted owl has shown us that. We have to stop going to the "blind men" for answers. I think the authors of *A Conservation Strategy for the Northern Spotted Owl* (Thomas et al., 1990) said this very well:

> We were asked to do a scientifically credible job of producing a conservation strategy for the northern spotted owl. We have done our best and are satisfied with our efforts. We have proposed. It is for others — agency administrators and elected officials and the people whom they serve — to dispose. That is the system prescribed in law. It seems to us a good one. We can live with that.

Frankly and bluntly, we must choose between (i) managing the forests for the values *we* place on them and (ii) walking away and allowing them to develop "naturally." Remember, Mother Nature is not always kind. Ask the wildlife that were either killed or driven out by the 500,000-acre Biscuit Fire. Ask the folks in New Orleans.

LITERATURE CITED

Franklin, J. F., K. Cromack, Jr., W. Denison, A. McKee, C. Maser, J. Sedell, F. Swanson, and G. Juday. 1981. *Ecological characteristics of old-growth Douglas-fir forests.* General Technical Report PNW-118. Portland, OR: USDA Forest Service, Pacific Northwest Forest and Range Experiment Station.

Franklin, J. F., F. Hall, W. Laudenslayer, C. Maser, J. Nunan, J. Poppino, C. J. Ralph, and T. Spies. 1986. *Interim definitions for old-growth Douglas-fir and mixed-conifer forests in the Pacific Northwest and California.* Research Note PNW-447. Portland, OR: USDA Forest Service, Pacific Northwest Forest and Range Experiment Station.

Jones, S. M. 1988. Old growth forests within the Piedmont of South Carolina. *Natural Areas Journal* 8(1):31–37.

Marcot, B. G., R. S. Holthausen, J. Teply, and W. D. Carrier. 1991. Old-growth inventories: Status, definitions, and visions for the future. In *Wildlife and vegetation of unmanaged Douglas-fir forests,* technical coordinators L. F. Ruggiero, K. B. Aubry, A. B. Carey, and M. Huff. General Technical Report PNW-GTR-285. Portland, OR: USDA Forest Service, Pacific Northwest Research Station.

Moeur, M., T. A. Spies, M. Hemstrom, J. R. Martin, J. Alegria, J. Browning, J. Cissel, W. B. Cohen, T. E. Demeo, S. Healey, and R. Warbington. 2005. *Northwest Forest Plan — The first 10 years (1994–2003): Status and trend of late-successional and old-growth forest.* General Technical Report PNW-GTR-646. Portland, OR: USDA Forest Service, Pacific Northwest Research Station.

Old-Growth Definition Task Group. 1986. *Interim definitions for old-growth Douglas-fir and mixed-conifer forests in the Pacific Northwest and California.* Research Note PNW-447. Portland, OR: USDA Forest Service, Pacific Northwest Research Station.

Potter, D., M. Smith, T. Beck, B. Kermeen, W. Hance, and S. Robertson. 1992. *Old-growth definitions/characteristics for eleven forest cover types.* San Francisco, CA: USDA Forest Service, Pacific Southwest Region.

Saxe, J. G. 1892. *The Poetical Works of John Godfrey Saxe.* Boston and New York: Houghton, Mifflin and Co., 1892.

Thomas, J. W., E. D. Forsman, J. B. Lint, E. C. Meslow, B. R. Noon, and J. Verner. 1990. *A conservation strategy for the northern spotted owl: A report of the Interagency Scientific Committee to address the conservation of the northern spotted owl.* USDA Forest Service, USDI Bureau of Land Management, Fish and Wildlife Service, and National Park Service, Portland, OR. Washington, D.C.: U.S. Government Printing Office.

U.S. Department of Agriculture, Forest Service. 1992. *Old growth definitions/characteristics for eleven forest cover types.* Internal memo. On file with: U.S. Department of Agriculture, Forest Service, Pacific Southwest Region, 630 Sansome Street, San Francisco, CA 94111.

U.S. Department of Agriculture, Forest Service. 1993. *Region 6 interim old growth definition[s] [for the] Douglas-fir series, grand fir/white fir series, interior Douglas-fir series, lodgepole pine series, Pacific silver fir series, ponderosa pine series, Port Orford cedar series, tanoak (redwood) series, western hemlock series.* Portland, OR: USDA Forest Service, Pacific Northwest Region.

U.S. Department of Agriculture, Forest Service. 2004. *Biscuit Fire recovery project final environmental impact statement.* Portland, OR.

White, D. L., and L. F. Thomas. 1995. Defining old growth: Implications for management. In *Proceedings of the eighth biennial southern silvicultural research conference,* compiled by M. B. Edwards. General Technical Report SRS-1. Asheville, NC: USDA Forest Service, Southern Research Station.

Wimberly, M. C., T. A. Spies, C. J. Long, and C. Whitlock. 2000. Simulating historical variability in the amount of old forests in the Oregon Coast Range. *Conservation Biology* 14(1):167–80.

Chapter 15

In the Shadow of the Cedars: Spiritual Values of Old-Growth Forests

KATHLEEN DEAN MOORE

Last on everyone's list of the values of old-growth forests—after goods; services; information; and cultural, recreational, and scenic values—is the sometimes vexing and usually undefined category of spiritual values. Yet old-growth forests hold particular value for the human spirit, the imagining and feeling part of the human mind. The challenge is to articulate, in secular terms, the sources of an ancient forest's spiritual values and their importance to human thriving. When the presence of spiritual values may be the best reason to conserve a forest, it will not do to be silent or confused about what those values are.

Ways to Value a Forest

For well over 200 years, Americans have valued forests as a means to the accumulation of personal or corporate wealth and the satisfaction of human material needs. A sign nailed to a sturdy Douglas-fir on an interpretive trail near my town identifies those values exactly:

Birth date:	1942
Diameter at breast height:	36″
Height:	124′
Board foot volume:	1,060 bf
Number of 8-foot 2 x 4s:	208
Rolls of toilet paper:	6,890

As children file by on their field trips, they learn to equate the value of a forest with the sale price of the material goods that can be extracted from the felled trees.

More recently, attention has been directed to another set of values, the ecological utility of the living forests—to shade streams, provide habitat, filter water, store carbon, modulate temperature variations and climate change, and on and on in a long and life-sustaining list of ecosystem services.

Then there are the cultural values that assess the usefulness of forests to human well-being, as that is understood to include more than the accumulation of wealth. Recreational and scenic values fall in this category—means to the ends of human health and happiness. Here we find also the spiritual values of forests, their power to lift and enliven the human spirit.

In the temperate rain forests of America's Pacific Northwest, near my home, the old-growth forests are places of spattered huckleberry light under fog-shrouded hemlocks, cedars, and Douglas-firs. These ancient groves speak with uncommon power to the imagining and feeling part of the human mind. They have the power to take a person's breath away, to make a person fall silent with wonder and gratitude, to deepen a person's connections to the wellsprings of life and death and mystery. All these qualities strengthen the spiritual well-being of anyone who walks in the shadow of ancient cedars.

Like its material and cultural values, the ancient forest's spiritual values can be understood as *instrumental* values. Instrumental value is the usefulness of the forest as a means to another's end—in this case, to the sense of spiritual well-being that humans seek. As philosophers point out, forests may also have *intrinsic* value. Intrinsic value is the worth of the forest as an end in itself.

But the presence of intrinsic value in the ancient forest is controversial. So, we will begin with a focus on instrumental value. What are the qualities of old-growth forests that most powerfully lift the human spirit? Or, put baldly, what explains the utility of old-growth forests for human spiritual thriving?

Spiritual Values of Old-Growth Forests
in the Pacific Northwest

Old-growth forests are old. At least in the Pacific Northwest, they are tall. They are complex. They are unspoiled. They are quiet. They are beautiful. They are all of these at once. These are their sources of spiritual value.

Continuity of Ages

In the old-growth forests of the H. J. Andrews Experimental Forest in the Oregon Cascades, hemlock saplings grow toward light shafting through cedars that have grown in this place for 500 years. The great age of the trees is impressive, but so is the vigor of the saplings, and what most impresses me is the gathering of generations, the unbroken span of ages throughout the forest. In contrast to the history-denying tree plantations of uniform age and simultaneous death, the ancient forest speaks of continuity, of the past that nourishes the present, and of the future that will grow from this ground in good time. Hundreds of years of history are written in tree rings, shelf fungi, broken limbs, crumbling soil, and sword-ferns shading the places where death and life cannot be distinguished.

Here is where a person can feel the deep history of the forest and the continuity of new life. There is comfort in this, to feel oneself, like the ancient trees, linked to past and future, part of the stream of living things in which nothing vanishes but grows from death into life again. In an old-growth forest, a person can find comfort in the immortality of substance and the constancy of change, can sink roots into the assurance that she came from the earth and will be folded back into it. There is value in the awareness of the continuity of life. And there is a cautionary wisdom, to understand that as the present creates what is to come, careless decisions now will shape only a precarious future.

Great Height

On a rainy day along Lookout Creek in the Andrews Forest, the ancient cedars reach into low clouds. Their mass looms solid and black and unbelievable. How trees can grow this big is beyond me. In fact, the whole forest is beyond me, and surely that is part of its spiritual value. "Man" is not the measure of these trees—our bodies are scarcely taller than the ferns that hide their boles, our arms too short to even approximate their girth. And how can we claim to understand this extravagant growth, when each new scientific discovery lengthens the ridgeline where science and mystery meet?

In a forest that has grown to this grandeur, a person comes into intimate contact with what is much more than human. The old-growth forest is graced by beauty we did not create, grown to heights we can scarcely fathom, and shaped by forces we cannot control, moving light and moisture and air on a scale we can barely measure. A forest of this height invites, maybe demands, humility.

It invites as well a sense of wonder—radical amazement at life on this majestic scale and in this microscopic exactitude. This often occasions an experience of the sublime, a mixture of fear and awe, to be in the presence of something that is both powerful and wonderful. In a secular world, stripped of all but material meaning, weary with worry, this encounter with the marvelous has great value.

I'm not necessarily talking about God. Many people feel closer to their gods in an old-growth forest. John Muir (1901) wrote, "The forests of America, however slighted by man, must have been a great delight to God; for they were the best he ever planted." But spiritual values should not be confused with religious values; not all spiritual people are religious, and not all religious people believe in God. Outside of the context of religious practice or belief, people respond spiritually when they yearn for intimacy with what is so wonderful in the secular world that it lifts their hearts. Given this, wandering quietly through an old-growth forest provides what many seek—a secular spiritual practice.

Complexity

No bulldozed, laser-leveled, monocultured, Roundup'd plots of genetically identical, age-identical trees arranged in rows can compare—an old-growth forest is complex, complete, and intricately interdependent. Ecologists speak of the structural diversity of the old-growth forest, its multilayered canopy, the shifting mosaic of windthrow and regrowth. They point to a diversity of species, life histories, populations, genetic information—from flying squirrels in their shredded-cedar dreys (nests) to mycorrhizal fungi among the roots.

An oversimplified forest plantation might allow a person to think that humans are managers apart from and in control of nature, able to create isolated and efficient systems. But an old-growth forest, with all of its parts in place and all of its systems engaged, demonstrates the fecund interconnectedness of soil, climate, water, life histories, and populations of plants and animals. The complexity of an old-growth forest invites us to ask how we fit into the earth's vast and complex cycles—not as managers, but as, in Aldo Leopold's words (1949), "plain citizens." Huckleberries or humans, our destinies tangle with the hunger of squirrels and the thirst of great cedars—interdependencies that are complex indeed and only barely understood.

This interconnection is important. In the intricate, life-giving relationships that link the destinies of people and places, we may find a stronger practical imperative to care for the systems that sustain us. In the complex relationships that fully create a forest, we may find a deeper and more complex vision of what it means to be fully human.

Tranquility

The old-growth forest trail in the Andrews Forest is the quietest place I have ever entered. When I sit on a fallen log to silence the noises my movements make, I can hear the flick of the stream across broken stone. Now and then, a vine maple leaf ticks against another, falling. But the forest hushes all the voices of the mundane world. Silence collects in the deep moss banks. From soaring arches, light pours down as if from leaded windows, casting an amber glow on the trunks of ancient trees and the buttressed stone. I hear a spotted owl hoot from a far place, although I can't be certain in this research forest if it is the owl itself or a biologist calling it in.

People call places like this "cathedral forests" for very good reason—the soaring architecture of draped stone and silence, the felt presence of something beyond human knowing, the sense of belonging to a community of celebration. But I believe there is an even more important meaning to the metaphor, and it is this.

In medieval England, when death was the penalty prescribed for every felony, a European could save himself by many forms of pardon, notably by taking sanctuary in one of the great cathedrals. No pursuer, no king's man, could harm him there. Cathedrals were profoundly powerful places of safety and transformation. Brought up short by a power greater than his own, the pursuer fell to his knees. Bathed in divine grace, the offender was cleansed of his offenses—reemerging as a new person, seeing the suddenly glorious world with new eyes.

When they are protected, old-growth forests are our sanctuaries, too. We come to the forest not as criminals but as fugitives nonetheless, fleeing the relentless pursuit of the phone, the drone of traffic and computers, the artificial light, the pace of personal ambition. In the ancient forest, we find peace and the chance to see the world again as if through new eyes—that fresh, that beautiful. The transforming power of sanctuary is the peace in a place that is protected from plunder.

"Natural" Condition

An old-growth forest is the outcome of natural succession, not the product of human artifice or the aftermath of anthropogenic destruction. It is, in that sense, natural, original, wild, and self-determined. Many define an ancient forest in negative terms: It has grown to a great ripeness because it has not been cut down or burned up or trampled or harvested.

On the trail that drops to the creek in the Andrews Forest, I can imagine myself following paths worn by the footfalls of people who have walked past these rough-barked trees for 500 years and have chosen not to destroy them. There *are* places in the world that no one has wrecked. An old-growth forest is living proof of the possibility of human restraint.

This is not a small thing. The presence of ancient, uncut forests provides a vision of what a forest can be if left to grow to its full beauty and ecological richness. Many people have never seen such a forest, mistaking tree plantations or third- or fourth-growth Douglas-fir stands for the real thing. Gradually, as the forests around us are stripped down and sold off, we accept the degraded as the norm—the way it's supposed to be, the way it's always been. Ecologists call this the sliding *ecological* baseline.

We acquiesce in a sliding *moral* baseline as well, asking less of ourselves as stewards of the land; expecting less of our leaders; excusing ourselves for the damage we do to beautiful, bird-graced places. And so we find ourselves in a time of a sliding baseline of *hope,* when all we can hope for is a compromise, a mitigation package, a delay in the destruction.

The fact of the old-growth forest blocks the sliding baselines—ecologically, morally, and spiritually. It bears witness: Here is what a forest biome can be; here is a measure of what we lose when we destroy a forest. Here is a measure of the enormity of the damage we have done in other places. At the same time, here is a testament to human restraint, in this place at least. And so finally, here is an example of what we might hope for: that we can save a place on earth and in our lives for unspoiled beauty. That we can come to understand that part of the great human project of learning to live on the land without wrecking it entails protecting some places that are original and wild.

Beauty

There is a line drawn on the land at the edge of the Andrews Forest. On one side lies a steep hillside of stumps, dirt bulldozed into heaps and ditches, and beside the road, the backseat of a car, riddled with bullet holes. On the other side, the uncut forest, filled with darkness at dusk, and the smell of warm firs. My friends and I sat on a stump big enough for all of us and wondered about beauty, another source of spiritual value in the ancient forest.

Beauty is a kind of wholeness, I believe, organic wholeness, as against what is severed and ugly. The beauty of an ancient forest begins with wholeness of sound (the humming creek, the harmonic whistle of the hermit thrush), of smell (that dark and dusky humus, chanterelles under cedar boughs), of sight (green sorrel leaves spread to the sky). Its wholeness, what Leopold called "integrity" is systematic, myriad functions growing into and from myriad forms. The astonishing whole calls for celebration, the way music calls to us—beyond thought, a direct awareness of worth, an unmediated gladness.

The forest's wholeness calls us to find within ourselves a wholeness of our own, a consistency of belief and action that we call "integrity" when we see it in a human being. If this is the way the world is, so astonishing and beautiful, then this is how I ought to act in the world—with gladness and celebration, with respect, with gratitude, with caring, that it might thrive. In this way, the beauty of the old-growth forest can call forth what is beautiful in us.

Intrinsic Value of Old-Growth Forests

This list of values takes us full circle, back to the initial distinction between instrumental and intrinsic value. The old-growth forest has instrumental values that fall in the category of spiritual values. I have explained six of them, six ways in which an old-growth forest does good things for the human spirit, that imagining and feeling part of the human mind.

But have my readers noticed this—that the forest has these instrumental values only because we believe it also has intrinsic value? Walking in an old-growth forest affects the human spirit in ways we judge useful, just because the forest is good beyond human hopes and larger than human aspirations. If the forest were only a commodity, if it had no meaning beyond its usefulness to human ends, would its descending light and sweet dampness set us back on our heels? Would we feel the presence of something beyond human perception? Would we find peace there, in the shadows and birdsong?

I submit that an old-growth forest has worth in itself, worth beyond human uses. It is a manifestation of the "fierce, green fire" (Leopold, 1949) of life growing across the face of the earth. Saving old-growth forests for their own sakes, for their intrinsic value as ancient communities of life, represents a novel moral achievement that goes beyond even the most sophisticated human self-interest.

When humans no longer inhabit the earth, when we no longer use the forest to satisfy our own needs, when we no longer celebrate its beauty, light will still spike through ancient forests, and moss will blanket hemlock roots. At each dawn, the ancient forests, will fill with what Robinson Jeffers (2001) called "the heart-breaking beauty that will remain when there is no heart to break for it." Even in the absence of humans, that will be a worthy thing. It is good for us to know that.

LITERATURE CITED

Jeffers, R. 2001. Powers and magic. In *The selected work of Robinson Jeffers*, edited by T. Hunt. Stanford, CA: Stanford University Press.
Leopold, A. 1949. *A Sand County almanac*. New York: Oxford University Press.
Muir, J. 1901. *Our national parks*. Boston: Houghton-Mifflin.

Chapter 16

Old Growth: Evolution of an Intractable Conflict

JULIA M. WONDOLLECK

The old-growth conflict has raged for over a quarter-century. It is a frustrating conflict for all involved, one that has engaged traditional adversaries such as environmentalists and loggers and nontraditional foes including schoolchildren and spiritualists. There have been innumerable efforts to try to resolve it, from a high-level forest summit involving the president of the United States to a malicious death threat made to the wholesome yet symbolic figures of Smokey Bear and Woodsy Owl. There have been lawsuits, administrative appeals, protests, and hundreds of hours of testimony offered at congressional hearings. Logging truck convoys have snaked through state capitals, trees have been spiked, and people have chained themselves to logging road gates and nested in majestic old growth. Passionate letters to the editor about all facets of the issue have filled the pages of local, regional, and national newspapers. The old-growth problem is a tangled web of questions, parties, science, and politics eluding simple solutions.

Discussions about old growth can at times seem like those of the proverbial blind men describing an elephant. Depending upon what piece of this immense creature one is examining, different observations can be made and conflicting conclusions reached. At one level, it exhibits all of the characteristics of an intractable conflict, one mired in philosophical differences, scientific ambiguity, and political symbolism. Yet, at another level, constructive collaboration among former adversaries is creating mutually acceptable management activities. How can a single issue manifest such divergent realities?

Roots of the Old-Growth Conflict

Conflicts arise when people care enough about an issue that it is worth their time and resources to engage in it. They are compelled by the magnitude of what they have at stake. People care passionately about old growth but for differing reasons. William Dietrich (1992) captured it well when he wrote, "The forest is the same, and yet different to each individual who looks at it. . . . [S]o many good people . . . love the forest so fiercely in such completely different ways." Agencies such as the U.S. Forest Service have an interest in maintaining authority over forest management decisions, public respect, and the sustainability of the resource for which they are responsible. Loggers have an interest in their livelihood and the well-being of their families. Formal nongovernmental organizations, whether advocating from an environmental or forest products perspective, have organizational interests at stake, including maintaining the support of a robust membership and protection of or access to the resource. The timber industry has an interest in its own long-term stability and the need to strategically plan operations with some level of predictability. Community leaders have an interest in the overall well-being of their communities, increasingly recognizing the connections between the health of human and natural ecosystems. Others have an interest in protecting the spiritual dimension of the old-growth resource and associated cultural practices. These are all legitimate interests, but they are divergent. Decisions that are obviously "right" to one group can be perceived as horribly "wrong" to another. And because what is at stake in old growth is so great, people are steadfast and vociferous in their advocacy.

Conflicts do not emerge in a void. Conflicts arise within a context that plays a significant role in defining issues and their boundaries, creating incentives, and triggering behaviors. The character of the old-growth conflict over time is a direct reflection of its social and political backdrop. It emerged in a time of increasing awareness and concern about the environment, activism on many environmental fronts, new laws providing new legal tools to try to influence decisions, and increasing distrust of government and government agencies. It was a time of growing concern about the environmental consequences of long-standing resource extraction practices. As the conflict took shape, people voiced sincere concerns about old growth, but they were also reacting more generally to dissatisfaction with government policies and practices in public resource management that were developed in a different time, imagining different issues, and with different social priorities.

Imperative for Collaboration, Reality of Conflict

Collaboration is essential to sustaining the viable presence of old growth in the landscape; it is essential to managing this conflict. No single individual, agency, or organization can independently accomplish this objective. Old-growth ecosystems do not recognize the geopolitical boundaries of national or state forests, tribal reservations, private ranches or farms, or municipalities. Old growth does not have a single home in our political system; many individuals, agencies, and governments carry some level of jurisdiction and authority over it. The science of old growth is similarly dispersed among federal and state agencies, universities, nongovernmental organizations, and private companies; no single organization is the sole repository of data and knowledge about it. More generalized knowledge about old growth also resides in communities that have historically used the resource, whether for cultural, economic, or spiritual reasons. The only way to bridge the jurisdictional boundaries, pool knowledge and expertise, and manage the old-growth resource at the appropriate scale is through people working together in a collaborative manner.

Old-growth conflicts play out on two stages with differing objectives and strategies. In the political arena, the broad policy debate is sustained and rancorous. On the community stage, however, the dynamic is more cooperative and collaborative, focused on problem solving. What accounts for these differences? How can conflict be raging on some dimensions of the old-growth issue while collaboration is occurring on others? Five maxims of dispute resolution highlight the distinguishing characteristics of disputes amenable to collaborative resolution and disputes that are not (Cormick, 1980; Susskind et al., 1999; Moore, 2003). Although these five maxims help explain why the old-growth policy conflict has proven so intractable (Lewicki et al., 2003), they also provide insight into why collaboration has emerged in places throughout the Pacific Northwest (Wondolleck and Yaffee, 2000). In these particular places, the larger, divisive policy questions regarding old growth have been taken off the table, and the problem at hand is bounded at a much more manageable scale. The primary focus of these collaborative processes is a landscape-level or site-specific problem with some degree of urgency for both the ecosystem and the associated community.

Maxim 1: People work together on difficult and controversial issues only when they perceive an incentive to do so. Parties to a dispute will collaborate in good faith only when they believe that collaboration provides the most promising path to meeting their interests. Their incentives will be weak if they do not perceive a sense of urgency to resolve the dispute or a need to work with the other parties to pursue their own interests. Although participants in the old-growth conflict may not be opposed to the idea of collaboration in general, those most concerned about influencing broad policy often believe that their interests are best served pursuing other strategic options, including filing lawsuits, mounting protests, and lobbying Congress for favorable legislation. Some see advantage in first shifting public opinion in their favor, working toward a change in political administration or representation, or supporting scientific research that they believe will bolster their position.

The swinging political pendulum of the old-growth issue over the past twenty-five years has muddied parties' incentives to collaborate, encouraging each group to believe that it has achieved a position of power and will prevail through the political process. When other strategic options exist that can empower a party's position in a conflict, there is little incentive to negotiate.

It is also worth noting that the old-growth issue has become entangled in a political process in which strategies and objectives at times have very little to do with old trees in a forest (Yaffee, 1994). For those engaged in the conflict because of the visibility it provides and the membership support and resources it generates through polarization, there is limited incentive to engage in processes that might resolve it.

In contrast, incentives differ markedly for those most concerned about specific places and pressing on-the-ground problems. Collaboration has emerged in places where there exists a sense of urgency about immediate problems, in particular forest restoration and community economic decline. The objective of those involved is to "solve the problem" as opposed to "advance an agenda." It is clear that many are tired of the conflict, recognize the legitimacy of each others' perspectives and concerns, and want to find new ways of dealing with the problem. As a participant in Oregon's Applegate Partnership put it, "We got to the point that there was something in it for everybody to start talking" (Wondolleck and Yaffee, 2000). Similarly, collaboration emerged involving the Gifford Pinchot National Forest after timber production was halted by timber sale appeals and the dense forest monoculture left from an earlier era of extensive clearcutting produced only poor-quality timber and degraded watersheds and wildlife habitat. Those who cared about this place formed what is now the Gifford

Pinchot Partnership. One partner commented, "Everyone was tired of the fight. Enough was at stake that people were willing to take risks to solve problems" (Little, 2005).

Maxim 2: Collaboration among parties to a dispute will not occur unless there is an opportunity to do so. When the number of parties to a dispute is small and communication among them easily facilitated, the parties themselves are often able to create legitimate opportunities for negotiation. Doing so is more challenging in disputes having a large and/or dispersed array of parties. The old-growth policy conflict does not reside in a single place with a clearly identified and singular authority. Old-growth forests can be found on state lands, federal lands, tribal lands, and private lands. Both the Bureau of Land Management and the Forest Service have responsibility for some old growth at the national level. Consequently, there is no single venue within which the debate about old-growth policy can be confined. Instead, there are multiple authorities, with varying missions, legislative mandates, and priorities, all of which care about what happens with old growth but none of which have the unilateral authority or power to resolve it.

In contrast, in places concerned with the well-being of a specific human community and its associated ecological system, people are more readily able to initiate dialogue and move beyond the broader policy ambiguity. The opportunities inherent in chance meetings on downtown streets or at kids' ballgames have been seized. Local leaders representing differing constituencies are taking the initiative to reach across battle lines and begin conversations that set the stage for collaborative problem solving. An informal gathering among agency, industry, community, and environmental group leaders on a backyard deck sparked the conversations that triggered formation of the Applegate Partnership. Those convened agreed that Saturday morning to work together to make "future land management in the Applegate Watershed ecologically credible, aesthetically acceptable, and economically viable." The stewardship contracting process has also provided critical opportunities, encouraging a focus on collaborative problem solving at the local level.

Maxim 3: Effective collaborative processes require the credible representation of affected interests. Representation of affected interests is a key attribute of effective dispute resolution processes. Credible representatives must be identifiable and able to participate. This maxim does not imply that everyone who cares about an issue must participate but rather that all who care about the issue are satisfied that their concerns are appropriately *represented* by someone within the process. These representatives must be respected and supported by those they represent and, hence, able to commit themselves and their constituencies to the implementation of any agreement. Without such representation, negotiations will not likely succeed, for the crack is left open for dissatisfied parties to say, "Wait a minute, I don't agree. I wasn't represented in that dialogue, and I dispute its conclusions." There is a staggering array of parties to the old-growth policy conflict, with diffuse and dispersed constituencies. It is very difficult to identify exactly who could credibly and legitimately represent whom in policy-level discussions, especially because many of the perceived "groupings" have multiple and sometimes conflicting perspectives. At the landscape or community scale, however, it is far easier to identify and engage credible and committed individuals representing an array of perspectives in a problem-solving mode. At this scale, it is possible to attain a level of transparency and accessibility of these processes, making it easier to keep interested individuals and groups both informed and involved and to reach out to new participants as the dialogue and issues evolve.

Maxim 4: People cannot solve a problem before agreeing on what the problem is. Some collaborative processes fail not because the right people were not at the table or because they were not bright and capable representatives of their constituencies but rather because each came to the table with a different definition of "the problem" to be solved and hence different criteria for creating and evaluating alternatives. This maxim suggests a strong challenge to resolving the old-growth policy conflict. Although significant advances in scientific understanding of old growth have occurred, the problem's boundaries remain ambiguous. The more we learn, the more truths of the past get set aside by new questions, the more assumptions and beliefs get revised. What exactly is old growth? Where is it? Why does it matter to ecosystems and society? What is needed to ensure its sustained presence and role in the forest ecosystem?

As highlighted throughout the chapters in this book, not only are there several definitions of old growth, but each definition also poses varied perceptions of the problem. If scientists are not able to agree on what constitutes old growth, where it currently exists, and what is necessary to ensure its sustained role in a functioning "healthy" ecosystem, then it is not surprising that the many parties to the old-growth conflict are in debate about it. To complicate the situation, science has also been used as a strategic tool in this conflict. The uncertainty associated with the science of old growth enables opponents to use science selectively to advocate for particular policy decisions, contributing to the confusion and complexity of the debate. When science is manipulated and politicized, its value and credibility diminishes in the public's eyes, making it all the more difficult to bound the problem in a widely acceptable manner.

In contrast, at the landscape and community scales the problems are quite apparent: degraded streams, eroding slopes, fragmented habitat, hazardous fuel loads, declining economies, stressed communities. These problems are immediate, visible, and defined. Their urgency prompts committed and serious engagement in efforts to find solutions. People are, in effect, saying, "While we may not agree on the exact characteristics, location, and appropriate policies regarding old growth in general, we nonetheless do agree that *this* particular site or landscape is in need of attention, and if *we* don't do something together and do it *now*, we will all lose in the long run." Collaboration does not magically make such problems easy to solve. As a partner in a Umatilla National Forest partnership put it, "At times it felt like we were just butting our heads up against each other; it seemed purposeless." However, the clarity and shared concern about these problems kept them working together, just as it has in other collaborative efforts.

Maxim 5: Issues involving fundamental value differences are not suited to collaborative resolution. Common ground is elusive in conflicts involving fundamental value differences. Although Gifford Pinchot and John Muir cared passionately about forests, their management philosophies diverged. Their conservation versus preservation divide was rooted in fundamental value differences that are undercurrent in the old-growth policy conflict today. Those who believe old-growth forests are sacred and no single tree should be felled are seldom able to negotiate management strategies with those who believe old-growth forests are wonderful and will remain so even if trees are harvested. Sometimes *perceptions* of value differences (i.e., "environmentalists" vs. "loggers") keep parties at odds even when their underlying shared interests may be unifying (Wondolleck et al., 2003).

At the landscape or community level, however, the question of concern is not who is right or whose values should prevail but rather how can those who care about the forest for whatever reason find ways to work together and solve problems despite their value differences? As one participant in the Gifford Pinchot Partnership put it, "Although the members hold many significantly different values, they have learned they can work together in the areas where their interests coincide" (Little, 2005). A logger in the Applegate Partnership summed it up when he said, "Once you can sit down and talk about a definable piece of land, you can get beyond philosophy, and things start to fall together" (Wondolleck and Yaffee, 2000).

Insights from the Old-Growth Conflict

Although understandably frustrating for all involved, the old-growth conflict has nonetheless served many important purposes. It has attracted widespread attention and illuminated the full array of values at stake. It has identified new questions and spotlighted the imperative for ongoing research. The impasse imposed by this conflict triggered a reprieve from long-standing management practices affecting old growth and a reassessment of future directions. Without the old-growth conflict, it is likely that there would be far less old growth in the forest today, and few parties would have been pleased with that outcome. In this sense, the old-growth conflict has been a positive conflict from a societal perspective (Coser, 1956) because it galvanized attention and concern, expanded the scope of scientific inquiry, stimulated new policies and plans, encouraged economic transitions in some communities, and enhanced public and scientific understanding of the old-growth ecosystem.

The old-growth conflict has not remained static. It has evolved in some fundamental ways, and the situation today differs from two decades ago. The boundaries and intensity of the conflict have shifted. Scientific understanding of the resource has expanded significantly. Some former adversaries are now working collaboratively on aspects of the issue rather than remaining at impasse. There is greater interagency communication and coordination today than there ever has been. Although no one has turned swords into plowshares, progress has nonetheless been made, and the swords are not brandished with the same frequency or fervor as in the past. This gradual evolution of the conflict has revealed some important realities about contemporary public resource management.

The old-growth conflict marks an important transition to a new era of public resource management. While ushering the National Forest Management Act of 1976 through Congress, the late Senator Hubert Humphrey hopefully suggested that it would "get the practice of forestry out of the courts and back to the forests" (Wondolleck, 1988). Former President Bill Clinton's Forest Summit was similarly cast as an effort to "bring peace to the forests of the Pacific Northwest." Neither promise was realized. Any public process today charged with allocating scarce and valued resources among divergent but legitimate interests is going to be a lightning rod for conflict. There is no way around this reality. The recommendations of an esteemed panel of scientists and the orders of Congress and the president of the United States were all unsuccessful at circumventing it. Even an edict from God that says "Thou shalt not harm old growth, and thou shalt ensure its sustained presence on the landscape" would not definitively settle this conflict. Although very few would disagree with the objective of such an edict, many would argue about the details of its implementation. What is "harm," and what and where is "old growth?" And what management actions will ensure its sustained presence? As revealed throughout the chapters in this book, these seemingly simple questions elude simple answers.

Management of public lands in the United States occurs in a context of conflict, and old growth has served as the poster child for this reality. Natural resource management today is about decision making in the face of complexity, uncertainty, and divergent but legitimate claims. In short, it is about managing conflict. It is worth noting that the only decisions that have been broadly supported in the domain of the old-growth conflict have been those generated by diverse parties learning and working together in credible and informed processes, focused on manageable pieces of the larger problem. Not only does this collaboration represent a notable evolution in the old-growth problem, one that certainly would not have been forecast in the early days of this conflict, but it also vividly demonstrates the imperative of collaboration to effective public natural resource management today.

LITERATURE CITED

Cormick, G. W. 1980. The "theory" and practice of environmental mediation. *Environmental Professional* 2(1):24–33.

Coser, L. A. 1956. *The functions of social conflict.* Glencoe, IL: Free Press.

Dietrich, W. 1992. *The final forest: The battle for the last great trees of the Pacific Northwest.* New York: Simon & Schuster.

Lewicki, R. J., B. Gray, and M. Elliott, eds. 2003. *Making sense of intractable environmental conflicts: Concepts and cases.* Washington, D.C.: Island Press.

Little, J. B. 2005. *Story profiles: Pinchot partnership.* Red Lodge Clearinghouse. http://www.redlodgeclearinghouse.org.

Moore, C. 2003. *The mediation process: Practical strategies for resolving conflict,* 3rd ed. San Francisco: Jossey-Bass.

Susskind, L., S. McKearnan, and J. Thomas-Larmer. 1999. *The consensus building handbook: A comprehensive guide to reaching agreement.* Thousand Oaks, CA: Sage.

Wondolleck, J. M. 1988. *Public lands conflict and resolution: Managing national forest disputes.* New York: Plenum.

Wondolleck, J. M., and S. L. Yaffee. 2000. *Making collaboration work: Lessons from innovation in natural resource management.* Washington, D.C.: Island Press.

Wondolleck, J., B. Gray, and T. Bryan. 2003. Us versus them: How identities and characterizations influence conflict. *Environmental Practice* 5(3):207–13.
Yaffee, S. L. 1994. *The wisdom of the spotted owl: Policy lessons for a new century.* Washington, D.C.: Island Press.

PART IV

The Challenge of Change—
New Worlds for Old Growth

Translating scientific knowledge and social views into land management actions requires a variety of social processes, including policymaking, economic valuation, adaptive management, and decision making. All of these pursuits occur increasingly on a global stage as wood is imported and exported along with ecological and social impacts. Part IV addresses the implications for the future management of old-growth forests from several perspectives.

Thomas, a former U.S. Forest Service research scientist and chief, characterizes the policy history of the issue and argues that the cumulative effects of policymaking over many decades have created a quagmire in which managers have little freedom to actively manage for restoration and other values. He argues for immediate actions to streamline cumbersome legislation. Stankey, a social scientist, observes that adaptive management is needed to deal with the uncertainties posed by the old-growth issues. He notes, however, that lack of trust has severely limited the potential of adaptive management to provide solutions for dealing with dynamic human and natural environments. Loomis, an economist, explains that many of the economic values of old-growth forests are not reflected in the marketplace; he claims that we need more explicit recognition of these ecosystem values if we are adequately to fund the public agencies that help provide them. Salwasser, a former Forest

Service official and now university dean of forestry, puts the Pacific Northwest in the context of global forestry today. With global demand for wood increasingly coming from plantations, including those of native species, the pressure to take wood from naturally regenerated forests is declining. But he argues that maintaining a globally competitive forest management sector in the Pacific Northwest region will help provide the industrial capacity to do needed restoration on older forests in the region. Finally, Lach, a sociologist, examines how decisions are made about complex scientific issues and provides some suggestions for how decision makers might act in the face of complexity and uncertainty.

Chapter 17

Increasing Difficulty of Active Management on National Forests— Problems and Solutions

JACK WARD THOMAS

To state my case plainly and simply: The governance of the public lands is broken. Immediate repair is in order.

The problem is clearly revealed in the management—or failure to manage—of federal old-growth forests in the Pacific Northwest. Beginning in 1891, a series of federal acts, each well intentioned, led to a quagmire from which the only exit is action to dovetail and streamline legislation. Along a trail littered with lawsuits, managerial adjustments, and the resulting hopelessly discouraged managers, old-growth forests have become cocooned in static management that is in keeping neither with the overall intent of applicable legislation nor with the Northwest Forest Plan (NWFP) developed to meet that legislative intent. This chapter considers how and why active management of the national forests has become ever more difficult, time-consuming, and expensive since 1990. It then offers a solution.

In the Beginning

The president was authorized to set aside forest reserves from the public domain by the Creative Act of 1891 (26 Stat. 1103: 16 U.S.C.A. 471). The Organic Administration Act of 1897 (30 Stat. 11: 16 U.S.C.A. 471) stated the purposes of the act as, "No national forest shall be established except to improve and protect the forest within the boundaries, or for the purpose of securing favorable conditions of water flows, and to furnish a continuous supply of timber."

Pressure from private timber producers assured that little timber was sold from the national forests from 1905 until the beginning of World War II in 1939. But, with the end of the war six years later, things changed. Little housing had been built between the onset of the Great Depression in 1929 and the end of World War II—a sixteen-year hiatus. Veterans returned from military service by the hundreds of thousands. A grateful nation greeted them with the GI Bill, which included no-down-payment, low-interest loans for home construction. As the timber supply from private lands was depleted, and public lands filled the gap (Hirt, 1994), the timber cut off federal lands gradually increased from less than 2 billion board feet per year (bbf/year) to over 13 bbf/year in the late 1980s.

The Multiple-Use Sustained-Yield Act of 1960 (16 U.S.C.A. 528-31) broadened U.S. Forest Service management responsibilities to include fish and wildlife, recreation, timber, grazing, and water. *Multiple use* was defined in this act as

> management of all the various renewable surface resources of the National Forests so that they are utilized in a combination that will best meet the needs of the American people . . . without impairment of the productivity of the land, with consideration being given to the relative values of the various resources, and not necessarily the combination of uses that will give the greatest dollar return or the greatest unit output.

Sustained yield was defined this way in the act: "the achievement and maintenance in perpetuity of a high-level annual or regular periodic output of the various renewable resources . . . without impairment of the productivity of the land."

Rise of Environmental Consciousness

Meanwhile, the Multiple-Use Sustained-Yield Act was mirrored in what has since been called a rise in *environmental consciousness* in the 1960s. The growing environmental movement pushed through laws in the 1970s that, coupled with myriad court decisions, changing public opinion, and availability of relatively cheap wood from overseas suppliers, reduced the Forest Service's timber program by more than eighty percent in the 1990s.

The plethora of laws that increasingly affected Forest Service operations began with the National Environmental Policy Act (NEPA) of 1970 (42 U.S.C.A. 4321-61). Over time, guided by court decisions, NEPA inexorably increased the time and costs of preparing assessments that preceded management actions. In turn, the Endangered Species Act (ESA) of 1973 (16 U.S.C.A. 1531-1543)—the "900–pound gorilla of the environmental laws"—had dramatic impacts on active management. The ESA also contained a clause, little noted at the time, that forced a dramatic change in land management: "The purposes . . . are to provide a means whereby the ecosystems upon which endangered species and threatened species depend may be conserved" (at 1531(b)). This was the precursor to *ecosystem management*.

Next, the Freedom of Information Act (1974) (5 U.S.C.A. 552) allowed the public to obtain detailed records of agency actions leading to agency decisions and facilitated challenges to agency activities. The National Forest Management Act (NFMA) of 1976 (16 U.S.C.A. 1600-1614) pursued three objectives: (i) allowing agency discretion to use even-aged timber management, (ii) mandating planning on a forest-by-forest basis, and (iii) ensuring that the Forest Service took the full range of resources into account in planning.

Building the "Perfect Storm"

The Forest Service took ten to fifteen years to prepare management plans for each national forest, as required under the NFMA, absorbing huge amounts of time and dollars—and not a single plan was ever executed as visualized and approved. The Forest Service proposed. The administrations, a string of congresses, the public, and the courts disposed (Hirt, 1994). New plans are now in preparation, but their future too remains in question.

Funding for timber harvests and associated road building was routinely approved, while requests for other activities involved in the forest plans were ignored or shortchanged. An *environmental deficit*—along with public backlash to the dominant emphasis on timber production at the expense of other factors—was building. Those with environmental concerns began to resort to court actions (Hirt, 1994).

The last piece necessary for the "perfect storm" relative to the Forest Service's timber program was the Equal Access to Justice Act (1980) (5 U.S.C.A. 504), which allows legal challenges by citizens who believe that proposed management actions are not in compliance with laws or regulations. Plaintiffs who win their cases are entitled to compensation for costs. If they lose, there is no penalty.

The Owl, the Scientists, and the Judges

The developing stalemate produced a "magic moment" that made it clear that a new course of action was required in management of public forests. By the late 1980s, the "perfect storm" was brewing over federal land management. The eye centered on the Pacific Northwest and the welfare of a cryptic subspecies of owl—the northern spotted owl. A master's thesis (Forsman, 1976) identified "old-growth forests" as the primary habitat of the owl. A flurry of research followed which, overall, sustained Forsman's conclusions (Thomas et al., 1990).

Elements of the environmental community seized on this information. They reasoned, correctly, that if the owl's primary habitat was old growth and that such forests on private lands had been logged and that logging was rapidly proceeding on federal land, then the owl was a candidate for listing under the ESA as "threatened." Almost certainly, there was already failure to comply with Forest Service regulations, issued under the auspices of the NFMA, to maintain viable populations of all native and desirable nonnative species, well distributed by planning area (Yaffee, 1994).

In truth, old growth was an icon, a symbol of wildness incarnate. The remnant stands of old huge trees were unique and had meaning and value to many that exceeded any value that might be derived from their being made into dimension lumber. It was the perfect battleground for those who valued such things. The owl was a symbol—and a lethal weapon against the status quo (Yaffee, 1994). The issue, in the parlance of the time, was "it's the old growth, stupid!"

Any action to save the owl (meaning old growth) would dramatically reduce timber harvest and cause significant job losses and likewise losses in revenues to county governments. But harvest continued above 4 bbf/year. Every effort to preserve even minimal amounts of old-growth forests in large enough patches to head off the listing of the northern spotted owl as "threatened" was successfully resisted until 1990 (Yaffee, 1994). In the meantime, President George H. W. Bush, at the Rio Earth Summit in Brazil, announced that, henceforth, national forests would practice "ecosystem management." Upon the listing of the owl as "threatened," that same requirement was effectively imposed by a federal judge.

Science Enters the Arena

Finally, the heads of the Forest Service, Bureau of Land Management, National Park Service, and U.S. Fish and Wildlife Service appointed a team of scientists to produce a plan to halt the decline of the spotted owl. The Interagency Scientific Committee to Address the Conservation of the Northern Spotted Owl (ISC) requested, and was granted, autonomy. The ISC's report (Thomas et al., 1990) produced a backlash from the administration, Congress, state governments, and all associated with the timber industry. Timber yield from public lands in the range of the owl would decline from 4.5 bbf/year to 2.5–3.0 bbf/year. Additional peer review of the report was ordered. Those reviews upheld the report. Then, in acrimonious hearings, the ISC was called to account in front of committees in both the Senate and House (Yaffee, 1994).

At first, the agency heads adopted the report. Then, Bureau of Land Management Director Cyrus Jamison pulled out of the agreement with promise of a better plan with less economic impact. Federal Judge William Dwyer lost patience and shut down harvest of old-growth timber until the government determined if Jamison's action rendered the ISC's strategy ineffective. Dwyer also presciently asked about the welfare of other species identified as associated with old growth (Yaffee, 1994).

In the meantime, the Forest Service chief assembled a separate Scientific Assessment Team to answer Judge Dwyer's questions. The team could not attain information as to what the "Jamison Strategy" entailed and told the judge so. They also determined that some hundreds of species, including mammals, birds, reptiles, amphibians, invertebrates, and plants, were associated with old growth (Thomas et al., 1993). Judge Dwyer continued his injunction.

Next, the ISC was put on public trial to establish a record for the consideration of the Endangered Species Committee (authorized in the ESA to review the costs of compliance and given the authority to allow a species to dwindle toward extinction if the costs were too high). To the embarrassment of the administration, the political appointees who made up the committee, including cabinet members, upheld the validity of the ISC report. Further actions were delayed until after the 1992 elections.

Over the next two years, two more committees of technical experts developed alternatives for consideration. In the end, every alternative fell back on the general principles and approaches put forward by the ISC.

Back to Politics

The House Agriculture Committee tried for a legislative solution. Chairman Kika de la Garza of Texas (encouraged by the Committee's chief of staff, James Lyons) recruited a group of scientists—the so-called "Gang of Four"—to prepare an array of alternatives to help assess impacts on employment and economic consequences (Johnson et al., 1991). In the end, the committee also chose to defer to the elections of 1992.

Presidential candidates Bush and Ross Perot promised adjustments in the ESA after the election. Candidate Bill Clinton said that he would, following election, hold a "Forest Summit" to hear all sides of the issue and then provide a strategy that would comply with the law(s) while being sensitive to the needs of those adversely affected. Bush and Perot split the conservative vote, and Clinton won Oregon and Washington with a minority of the votes—and the national election.

Almost immediately, Clinton ordered a new effort to formulate a solution. Staff came from land management and regulatory agencies, state agencies, academia, and the private sector. Recommendations were to be confined to federal lands. There was to be an array of feasible alternatives and an assessment of economic, ecological, social, and legal costs, all to be developed under the concept of ecosystem management. The definition of "ecosystem management" was left to the discretion of the Forest Ecosystem Management Team (FEMAT) that Clinton created. The report was due in ninety days.

Clinton chose "option 9" out of ten options presented, which was to become, after significant modification, the NWFP. Some ninety percent of old growth on federal lands was placed into protected status. Timber harvest was projected to be 1.25 bbf/year. The FEMAT identified more than 1,000 species that were potentially associated with old-growth ecosystems. Several hundred were considered "at risk," primarily because so little was known of their habitat requirements (FEMAT, 1993).

Other teams prepared the assessments and documents required by the NEPA and the ESA. Advising legal staff demanded attention to viability and distribution of all species thought to be associated with old-growth forests. "Survey and Manage" protocols were imposed that, prior to any management action in old growth, required a survey for some 400 species which, if found, required mitigation in any proposed sales. Timber yield projections dropped from the 1.2 bbf/year projected in the FEMAT report to 1.1 bbf/year, which was subsequently revealed as a gross overestimate.

Original FEMAT members pointed out that the Survey and Manage protocols would prove extremely expensive and would render any harvest of old growth theoretically allowed in the plan extremely difficult and most likely impossible. That advice was ignored, although it proved correct in the long run.

Survey and Manage turned the ESA's requirements upside down. The act required an elaborate process to declare a species "threatened" or "endangered," but the Survey and Manage protocols, in essence, assumed species in question to be "threatened" until proven otherwise. This was what made any harvest of old growth inordinately expensive and highly unlikely. Subsequent failures to use the full flexibility of "adaptive management areas," coupled with failure to adjust riparian buffers as specified in FEMAT, resulted in more reductions in active management for any purpose. Timber harvested has dropped as low as 250 million board feet/year, as opposed to a projected 1.1 bbf/year.

Ten Years Later—How Are We Doing?

A ten-year review of efficacy of the NWFP in 2006 indicated success in meeting the biological/ecological objectives for old-growth forests and dependent species. However, it had failed to meet projected timber yields and had fallen short in dealing with displaced workers and isolated communities that had been almost totally dependent on the timber industry.

The same scenario, for similar political reasons but with different ecological scenarios, reduced timber yields from national forests across the United States by similar percentages over the 1990s. Nationally, timber harvest dropped from 13 bbf/year in the late 1980s to less than 2 bbf/year in the early 2000s, a reduction of more than eighty percent. Other activities such as thinning to reduce fire risks have been difficult to execute in an expeditious and efficient manner.

President George W. Bush's administration produced the Healthy Forests Initiative and guided the passage of the Healthy Forest Restoration Act (HFRA) in 2003—the first legislation to mention old growth. The planning rules prepared under the auspices of the NFMA have been revised, and a task force has been established to consider revisions to the NEPA. President Bush appointed Mark Rey, a former top-level timber industry lobbyist and former chief of staff to the Senate Natural Resources Committee, as under secretary for natural resources and environment, with responsibility for the Forest Service. Nonetheless, timber harvests have remained at the same low levels for the six years since, in part due to an atmosphere riddled by suspicion.

President Bush took office backed by a Republican majority in both the Senate and the House. The stars seemed aligned for enhanced timber yields from the national forests. What happened? Some management energy was diverted to reducing fuels in the wildland–urban interface. But, all in all, active forest management increased only marginally. It can be concluded that it really matters little for national forest management who occupies the White House and controls the Senate and the House. The real problem is probably more systemic than political.

The perfect storm gains strength with each court decision lost by land management agencies. The operative laws, although individually rational and appropriate, mesh ever more poorly with a stream of judicial decisions that "clarify" (or confuse) just what the laws add up to in varying combinations. Documents that were once less than fifty pages now run into the hundreds of pages.

Appellants, usually of the environmental persuasion, are increasingly sophisticated in opposing projects with which they disagree. Appeals and suits are commonly executed with astute timing that causes maximum cost and disruption to the agency. The first step is to request an extension in the comment period, which if granted, costs time—sometimes a full field season. First appeals are most commonly based on alleged failure to follow appropriate processes. If success garners an injunction and a delay in operations, it can and often does result in "victory"—in that action is delayed, the agency distracted and, often, time-sensitive actions are no longer economically viable. If that appeal is rejected, a suit is filed against the planned action as a violation of one law or another, and an injunction is requested pending trial. Win or lose, injunctions can delay projects, and a trial, win or lose, eats up more time and increases costs for the government. In the end, win or lose, the proposed action will cost more, sometimes much more, than any benefit *or* detriment that might have resulted from said action.

Morally Bankrupt, Economically Irresponsible, and Socially Careless

There is no accepted calculus for measuring public opinion. So, when has "public input" been adequately considered? Consider the controversy on the Bitterroot National Forest in Montana relative to legal challenges to proposed timber salvage and recovery activities in the aftermath of wildfire in 2000. Counting up the "mail" (much in the form of preprinted post-cards and much from far away) clearly indicated that substantive salvage actions were unacceptable to several people. However, elected officials—two senators, the one congressman, the governor, local county commissioners, and state senators and representatives representing the areas involved—unanimously pushed for quick and extensive salvage. How does "democracy" come to bear in such circumstances?

Administrations, regardless of party in power, have, since 1945, rather consistently pushed land management agencies to meet "timber targets." "Environmentalists" have increasingly opposed most active management that has a commercial logging component, although there is some ambivalence about reducing fire danger in the wildland–urban interface as encouraged by the HFRA. Presumably, opposition to making houses and their occupants safer from wildfire would be "politically incorrect."

This ever-increasing snafu is well recognized and aging news. Additions to extant laws will likely further confuse and upset the increasingly dysfunctional status quo. The situation is apt to continue and intensify, unless Congress and the president step in.

We must exploit our environment to live. The operative question is how to do this in an acceptable and sustainable fashion. If any nation is capable of actively managing forests, including public forests, to produce an array of goods and services, it surely is the United States. What's more, we can probably do this while accommodating the traditional multiple uses, sustaining the evolving attributes of "ecological services," and rendering forests less susceptible to stand replacement fires. As use of this nation's resources decreases, demands increase on the resources of other nations—most of which are not as well equipped in technical expertise and financial resources to manage forests sustainably. In other words, our rapacious environmental impacts are exported. At the same time, dollars are exported (in the face of a rapidly burgeoning negative balance of trade and spiraling national debt) along with jobs associated with active management. This combination of results is, plain and simple, morally bankrupt, economically irrational, and socially deficient.

Draining the Swamp

The president and the Congress need to appoint a new Public Land Law Review Commission. Past commissions have neither clarified the cumulative intent of the law(s) nor streamlined required processes and procedures. Why? The situation had not yet reached an apparent crisis point. Instead, they produced voluminous reports that, in the end, stimulated some relatively minor actions but left looming issues unaddressed. The reports were long on rhetoric and short in terms of significant reform. And, perhaps they had too much money and too much time.

A new commission could be composed of truly knowledgeable and experienced individuals. The task (and the charge) should be simple and direct: The system is broken, and you have some definite period to produce several alternative packages of revisions in the law(s) that will clarify missions, streamline operations, eliminate overlaps in responsibility, clarify how "public opinion" will be handled in natural resources planning and execution, ensure oversight, and minimize court involvement.

These suggestions might overcome previous commissions' shortcomings:

- There should be a limited time for execution—say six months. The report should be delivered to Congress and the president at the beginning of a new Congress so as to be as sheltered as possible from the members' second-year fascination with upcoming elections.
- The key members should be expected to work full-time on the project.
- Commission members should be compensated at the rate of the highest level of the senior executive service.
- Adequate support staff should be made available as requested by the chairperson.
- The effort should begin with the full recognition that there are problems that will require adjustments in law.
- Results should be largely in the form of several alternative courses of action, packaged as legislation ready for introduction.
- Clarity of purpose, intent, and required process should be of paramount concern.
- Efficiency of management (in both time and money) should be a primary concern.
- Consideration should be given to establishment of an arbitration process to handle disputes with agency actions short of federal court.
- Rights of citizen participation in planning and rights to appeal should be preserved. However, it is imperative that processes be

instituted that prevent "game playing" to draw out decisions and impose costs.

- It should be recognized that the existing panoply of laws, as construed by the courts over the years, has created an effective (if quite burdensome, cumbersome, and inefficient) system of accountability that checks both agencies and Congress. The outcome of deliberations should include some similar level of accountability but with streamlined processes that reduce their ponderousness.

Action Is Overdue

Congress and the president must act. To continue down the current path is wasteful, destructive, unaffordable, and divisive and presents a compounding message that the government is impotent—or simply does not care—to rationally address increasingly obvious problems. If there is a political issue that cries out for, and deserves, bipartisan cooperation, it is the sustenance and management of our public lands. The governance of our public lands is "flat dabbed broke," to put it mildly. More patches on top of older patches in the form of "add-on" legislation will not fix the problem.

But, will Congress and the president act? A cynic, with good reason, might be doubtful. Time will tell.

In the process of facing the issue of old-growth forests, we learned much about the ecological and political complexities of managing public forests. That ongoing saga continues to inform and teach us about the social, political, and legal aspects of the complexities at hand, should we care to learn. We are still, thirty years after the old-growth issue came to the forefront, wading through the swamp, first in one direction and then in another. Clearly, our maps are faulty and our compasses broken. This issue demands and deserves bipartisan cooperation to draw a new map and calibrate our compass to guide public land management in the twenty-first century.

LITERATURE CITED

FEMAT (Forest Ecosystem Management Assessment Team). 1993. *Forest ecosystem management: An ecological, economic, and social assessment*. Report of the Forest Ecosystem Management Assessment Team, 1993–793–071. Washington, D.C.: U.S. Government Printing Office.

Forsman, E. D. 1976. *A preliminary investigation of the spotted owl in Oregon.* M.S. thesis, Oregon State University, Corvallis.

Hirt, P. W. 1994. *A conspiracy of optimism: Management of the national forests since World War Two.* Lincoln: University of Nebraska.

Johnson, K. N., J. F. Franklin, J. W. Thomas, and J. Gordon. 1991. *Alternatives for management of late-successional forests of the Pacific Northwest.* A Report to the Agriculture Committee and the Merchant Marine Committee of the U.S. House of Representatives.

Thomas, J. W., E. D. Forsman, J. B. Lint, E. C. Meslow, B. R. Noon, and J. Verner. 1990. *A conservation strategy for the northern spotted owl: A report of the Interagency Scientific Committee to address the conservation of the northern spotted owl.* USDA Forest Service; USDI Bureau of Land Management, Fish and Wildlife Service, and National Park Service, Portland, OR. Washington, D.C.: U.S. Government Printing Office.

Thomas, J. W., M. G. Raphael, R. G. Anthony, E. D. Forsman, A. G.Gunderson, R. S. Holthausen, B. G. Marcot, G. H. Reeves, J. R. Sedell, and D. M. Solis. 1993. *Viability assessments and management considerations for species associated with late-successional and old-growth forests of the Pacific Northwest.* Portland, OR: USDA Forest Service.

Yaffee, S. L. 1994. *The wisdom of the spotted owl: Policy lessons for a new century.* Washington, D.C.: Island Press.

Chapter 18

Is Adaptive Management Too Risky for Old-Growth Forests?

GEORGE H. STANKEY

The media today are full of stories about the old-growth "problem." But, we might ask, what exactly is the problem? For some, there is a strong conviction that the nation's old-growth forests have been all but lost. For these people, like the MasterCard ad says, old growth is priceless. As the source of a host of scientific, ecological, spiritual, and cultural values, such forests must be protected. Others perceive old-growth forests as part of the wider fabric of nature, evolving over time, even disappearing as a result of natural and human events (fires, logging), then reappearing elsewhere. In this worldview, old-growth forests are the product of both ecological processes and human choices. If more old growth is desirable, then actions can be taken to achieve this end.

Although this summary overly simplifies the complexity, intensity, and diversity of arguments about old growth, it does reveal an important characterization of old-growth forests *as a problem*. It is predominantly a two-faceted debate: a debate about the science of the ecological processes that give rise to such forests *and* a debate grounded in deep, often divisive, value systems. Old growth shares commonalities with many contemporary resource problems, such as the importance of endangered species and biological diversity.

Such problems have been described as *wicked*. This term has come to mean that they are seemingly straightforward technical challenges, but the underlying values involved turn them into complex issues whose satisfactory resolution is elusive. It has been argued that such problems have no right or wrong answers, only more- or less-effective solutions. For such problems, science can play an important role in informing the debate—helping people understand the underlying processes involved, the functions such systems offer, the consequences and implications of alterations to the system—but answers inevitably play out in the political realm, not the scientific.

Such problems, and the conflict and contention they evoke, increasingly characterize forest management, both in the United States and globally. When I began my career in 1968, debates over clearcutting on the Bitterroot National Forest, and particularly the practice of *terracing*—leveling and scarifying steep slopes to enhance seedling survival—already were aflame. Eventually, this controversy led to a congressionally mandated review of forest management practices and also to both the Forest and Rangeland Renewable Resources Planning Act (RPA) of 1974 and the National Forest Management Act (NFMA) of 1976.

In 1993, forest management conflicts brought the president of the United States to Portland, Oregon, to find a solution to the "gridlock" gripping the region. Many issues were involved: the future of rural timber-dependent communities; the survival of a host of forest-dependent species; and, in particular, the need to ensure the protection and enhancement of the region's old-growth forests and associated species and values. Indeed, it was the iconic image of old-growth forests or, as they were dubbed by environmental organizations—*ancient forests*—that largely catalyzed public concern. Big trees, evoking feelings of awe, spirituality, and humility, became the central representation of these public concerns. How the existing stock of old growth, greatly diminished over the past century, could be protected and how more might be created were central questions that the Forest Ecosystem Management Assessment Team (FEMAT), created by President Bill Clinton in early 1993, sought to address.

Defining *Old Growth*

As other chapters in this volume note, *old growth* does not define easily. When exactly does a forest become "old growth?" Although the scientific literature is divided on this issue, generally we are talking about trees whose age exceeds that of most humans—at least eighty years and typically much older. There is also increasing agreement among scientists that the structural characteristics of old-growth forest, as opposed to simply their age, are the most useful defining elements. Thus, we consider attributes such as multi-layered canopies, standing and down dead trees, varying tree sizes and ages, and so forth (Spies, 2004), although we remain also attached to the ideas of minimal human influence and long time periods. It is especially important to identify structural qualities of old-growth forests because it is these attributes that offer insight as to how the extent of old-growth forests in the future might be increased. In other words, it might be possible to manage forest ecosystems to accelerate creation of old-growth conditions in ways other than simply letting time go by or by establishing "no human intervention" reserves. However, the questions of what specific strategies are possible and how they might be implemented are less clear and address whether any single approach across all landscapes and ownerships makes sense.

Shall We Adapt?

One approach to addressing such challenges that has gained currency in recent years is *adaptive management*. Adaptive management has attracted serious scientific attention for about thirty years. At its core is a simple idea: "Management policies are experiments—learn from them!" (Lee, 1993). Thus, adaptive management is promising because it helps address the concern about the frustrating rate at which knowledge is acquired through traditional scientific inquiry.

Science typically breaks problems down into smaller components to better understand underlying functions and processes. However, such knowledge, although a part of the information base needed to manage complex ecosystems such as old-growth forests, is insufficient to address issues that involve multiple, integrated sectors, resources, and values. Adaptive management proposes that policies be designed in such a way that they maximize the learning associated with implementation, thereby allowing land managers and society to modify actions in light of any new knowledge that emerges. In this sense, adaptive management attempts to mimic the scientific method—although focused on questions that are more integrated and ultimately more complex than in typical research studies—by comparing alternative land management practices and policies.

It's interesting to note that the first academic discussions of adaptive management portrayed it as a technical undertaking, with particular attention to such issues as statistical design and analytic rigor (Walters, 1986). However, over time, there has been growing recognition that adaptive management is as much a social and political undertaking as it is scientific. Kai Lee advocated the idea of *civic science* as a crucial component of adaptive management—citizens both contribute their knowledge to adaptive processes and are informed by the scientific community in ways that ensure sound decisions. The result is decisions implemented through civic and informed discourse. Under this conception, adaptive management has a potential to contribute to both the technical and the social/political components of the old-growth problem.

Adaptive Management in the Northwest Forest Plan

Adaptive management is a central part of the Northwest Forest Plan. The plan acknowledged that, in the short term, with species at risk and limited understanding of how various forest management prescriptions might affect such species—positively or negatively—a conservative approach was warranted. However, in the longer term, as new knowledge and understanding emerged, policies and allocations in the plan would be subject to change. Thus, adaptive management would be the *engine* that drove the plan's evolution over time.

The plan employs adaptive management in two ways. As suggested above, it was to be a central implementation strategy throughout the plan area; an adaptive strategy, employed over time, would permit the plan to evolve in response to emerging knowledge. In addition, to ensure latitude for the policy experimentation an adaptive approach would require, the plan allocated about six percent of the region to ten Adaptive Management Areas (AMAs). Within the AMAs, managers, scientists, and citizens were encouraged to seek technical and social innovations in land management practices and institutional links among various players.

Given the complexity and uncertainty of framing management policies and actions that not only would sustain, but help create old-growth forests, an adaptive approach seems well suited. As discussed in the contemporary literature, adaptive management involves explicit discussion of the questions to be addressed (what are we trying to discover?), statements of hypotheses underlying various policies (what do we expect as a way to gauge success?), documented procedures (what exactly will be done?), monitoring protocols (what are the results?), and open interpretation and evaluation processes (what does it mean?). Such an approach would foster two types of knowledge: that needed to sustain and increase old growth as well as that needed to evaluate the efficacy of alternative management approaches.

Failing to Adapt

But efforts to apply an adaptive approach to a variety of resource management situations, including the Northwest Forest Plan, reveal a largely disappointing story. A recent literature review of adaptive management, for example, reported that, despite the potential and promise of the strategy, results have proven modest at best (Stankey et al., 2005). This seems the case across a variety of resource management sectors (e.g., agriculture, fisheries) and sociopolitical settings. Various factors account for this.

Perhaps the most fundamental challenge lies in the term *adaptive*. A defining characteristic of human behavior is its adaptive quality, grounded in learning. We learn from experiences: For example, don't touch a hot stove. But such learning comes less from any formal process of specifying a question; undertaking some type of experiment; and formally evaluating results than it does from simply acquiring information, sometimes painfully, that "B" follows "A." However, in the case of adaptive management, successful implementation has been handicapped by a tendency to assume "we've always been adaptive." Although it might be argued that the plan itself represented an effort to adapt to a changing world, the strategy of promoting a formal program of adaptive management was judged by many to reaffirm simply what management agencies have always done. And, if this is the case, then there is little incentive or perceived need to think critically about what kinds of management structures and processes, skills and abilities, or incentives and disincentives are needed to create an adaptive organization.

The situation is confounded by the generally risk-averse nature of contemporary society. Avoiding risk, with all its unknown dangers, has come to dominate much thinking, not just in natural resource management. The plan is grounded firmly in a risk-averse strategy: Effectively, it notes past mistakes, errors, and poor judgments and asks, why jeopardize the futures of species and conditions, such as old growth, that we hold valuable and important by undertaking actions for which outcomes are unknown? For many managers, there is little incentive (and many disincentives) to embark upon what others might consider risky undertakings, and much of the legal milieu under which they operate reinforces such behavior.

In the case of the plan, much management activity is governed by a host of standards and guidelines (S&Gs) that provide formulaic rules and procedures to which managers are held accountable. Such a rule-based approach to management stands in sharp contrast to an adaptive approach grounded in localized knowledge and management actions geared to idiosyncratic conditions and contexts, both social and ecological. Ironically, this was the original intent behind creation of the AMAs—to provide locations where innovative ideas could be pursued and where managers would have a license to operate outside the S&Gs. This has proven difficult, and recent policy decisions have discouraged such "nonconforming" behaviors in favor of requiring actions consistent with the S&Gs.

Because adaptive management explicitly acknowledges a lack of knowledge and proposes policy implementation as a means of acquiring new understanding, there is inevitable uncertainty in how things will work out. Of course, much of life is uncertain, but we manage to function. However, when something of value, such as old growth, is involved, and the fragile status of that resource is judged—rightly or wrongly—to be the result of prior policies and programs, it is not surprising that there is a reluctance to experiment. This is exacerbated when the issue is characterized by strongly held values, high scientific complexity, and an absence of trust.

Society in general, and resource management agencies in particular, have witnessed a steady diminution in trust in recent years. One result of this is an increased reliance upon legal and rule-based decision-making processes (despite the basis of many of these rule-based prescriptions on limited and incomplete understanding) and a parallel reluctance to delegate discretion to local resource managers. Another way to put this is that the burden of proof has shifted. Once, actions could be undertaken until evidence revealed undesirable results that needed to be terminated or modified. Now, unless compelling evidence exists regarding the likely outcome of a policy, many innovative management actions lack the social acceptance needed to initiate them. Adaptive management, with its aggressive stance with regard to uncertainty, will be seen by some as too risky to undertake, particularly when highly valued resources are at stake. Trust, or the lack thereof, becomes especially important when uncertainty is high. In the past, science was marked by high levels of social trust, but today, scientists find themselves subject to the same kinds of skepticism other practitioners face.

We must also acknowledge that some groups consider efforts to employ adaptive management to enhance and expand old-growth forests as inimical to their interests. In the case of the plan, about eighty percent of the twenty-four-million-acre region is protected as some form of reserve, where human interventions, such as road building and timber harvesting, are largely prohibited. From one perspective, this would suggest that old-growth forests and associated species and assemblages have gained significant protection.

Why, some might ask, give up these achievements to an "adaptive" policy that could reduce these protections for a possible, but uncertain, outcome in the form of more old-growth forests?

In this sense, an aggressive, active adaptive management program can be seen as threatening. For example, there is growing scientific evidence that thinning, prescribed burning, or other active interventions could enhance and/or speed development of the structural characteristics of old growth, reduce risk of loss to high severity fire, or maintain forest types dependent on frequent (but less-severe) fires. However, for some interests, such knowledge could foster arguments that old-growth reserves were unnecessary and could be eliminated or reduced in size. For many, this would constitute a step backward: The (presumed) protection offered by reserve status would be replaced by a strategy in which existing old growth could be harvested on the grounds that emerging knowledge about forest dynamics makes it possible to create a never-ending supply of old growth. However, in an environment laden with distrust about resource managers and scientists, it is easy to see how this assertion would always remain suspect.

Uncertain Challenge of Learning

The core motivation of adaptive management is *learning,* not as an end in itself but as a means of informing subsequent action. Yet learning is often resisted. Learning can reveal that existing policies are not going to lead to the outcomes expected and desired. This means change is needed, but such change often faces resistance. The comfort and security of doing things as they've always been done can outweigh the risks and uncertainty of doing things differently. Learning can also mean that policies and programs that serve one's interests might be replaced by others less favorable to those interests, or at least perceived to be so. In either case, despite the general belief that learning is a good thing, there can be strong opposition. "[M]ost people under most circumstances are not all that eager to learn. . . . Most . . . are content with doing things as they have always been done" (Michael, 1995).

But without learning and a commitment to integrate new knowledge into informed policymaking processes, the notion of adaptive management becomes empty rhetoric. Lacking commitment and substance, it risks joining the ranks of similar proclamations that litter the resource management landscape, such as multiple use and ecosystem management. Is this to be its fate?

Significant stakes are involved here. If active management advocates are right with regard to old-growth management, but society chooses, for whatever reason, not to adopt an aggressive adaptive approach, we could expose these forests to even greater risks than they currently face. Let's consider the variety of real risks associated with policies grounded in nonintervention. Old-growth stands could be lost to high-severity fires, windstorms, and insect and disease outbreaks. Some old-growth types, such as open ponderosa pine, completely disappear as they fill in with shade-tolerant species and then are consumed by intense fires. Climate change, that pervasive but poorly understood global force, could alter growth rates, establishment processes, and disturbance frequency and severity. Any of these outcomes are possible when uncertainty prevails, particularly under the current regime of reserves. In other words, drawing a line around such areas and prohibiting any human intervention could ultimately prove detrimental to the old-growth resource.

At the same time, we have to acknowledge that even if we succeed in implementing an adaptive-based approach to old-growth management, it could eventually fail; conditions could worsen; and the resource, despite our best efforts, could be lost.

Breathing Life into Adaptive Management

So, what's needed to give substance to adaptive management, to make it a viable strategy to be employed in the stewardship of old growth? Two ideas seem important.

First, there is little doubt that, unless we can foster trust among interested parties—citizens, scientists, policymakers, and managers—there is little chance that adaptive management can succeed. Trust is critical in situations in which there is a state of risk and a need for one party to depend upon the actions of another. As discussed earlier, adaptive management is a strategy based on dealing with both the unknown and with risk. It follows that social license (on the part of the wider community) is required to support action (by resource professionals) in the absence of understanding. When one understands the source of risk (danger), it is rational to avoid that risk. However, if action in the face of the unknown must be accompanied by an assurance that nothing will go wrong, then we have a recipe for inaction. But inaction *is* an action, and as noted, it is one that could jeopardize the future of old growth as much as any unfounded intervention.

When uncertainty and complexity are high, an adaptive management approach calls for probing the unknown—cleverly, judiciously, and with grounding in the best available knowledge—but also acknowledging that we do not know exactly what is going to happen. The process of framing problems adequately and identifying key hypotheses and questions should reveal likely outcomes, but inevitably, uncertainty will remain. For an adaptive approach to work, trust and mutual respect among the various parties must exist, and participants must acknowledge the provisional nature of their actions.

Unfortunately, there is a reluctance to engage in this behavior or even to allow it to occur. An innovative contribution of the Northwest Forest Plan was creation of the AMAs, where precisely such testing, probing, and experimentation could occur. The AMAs were to serve not only as outdoor laboratories but also as places where interested parties could learn to work together, where mutual respect and trust could be fashioned, and where informed policy could take root. In this sense, the AMAs had the potential to become *venues for working through,* places where interested parties could learn; educate; and act in a collaborative, collegial manner (Stankey and Shindler, 1997). Some impressive results have come from the AMA program. The Applegate AMA, near Ashland, Oregon, became a model of successful collaboration between citizens and managers. Work at the Central Cascades AMA, east of Eugene, Oregon, demonstrated how interactions between scientists and managers could reveal creative approaches to land management policies. Despite these and other important gains, however, the AMA program was dropped as a funding priority from the plan in 1998, only four years after it was signed.

The second idea important to making adaptive management viable is that using it to frame informed public policy for old-growth management (or any other forest management issue) requires that we acknowledge the limits of science and technical knowledge. These are necessary but not sufficient qualities of enlightened decision making. A systemic problem facing contemporary resource management policymaking is what some refer to as the *technical information quandary.* Society faces many problems that are complex and deeply grounded in technical disciplines; however, in a democratic society, effective policy requires public understanding and support—policy must possess social license. The era of the trusted expert whose pronouncements and interpretations are accepted as "truth" is gone. Yet their expertise remains important. The challenge is to find processes through which that expertise can be incorporated into political and social processes in ways that inform, educate, and modify policy.

Science is critical to questions of *what*, *why*, and *how*, but it is only one input to questions of *should*.

The challenge of employing an adaptive approach to old-growth forest management is primarily social and institutional in nature. Its successful promulgation requires systemic changes. Specifically, it requires creating an environment in which uncertainty is neither ignored nor denied but seized, where innovation and risk-taking are encouraged and rewarded, and where learning replaces the search for blame.

At this point, our capacity to create such an environment is questionable. For the future of old growth, this is not encouraging. In other words, the new world for old growth is one of a shifting burden of proof and increasing risk aversion. Ironically, such social trends risk standing in the way of achieving old-growth and forest conservation goals in the longer run.

LITERATURE CITED

Lee, K. N. 1993. *Compass and gyroscope: Integrating science and politics for the environment*. Washington, D.C.: Island Press.

Michael, D. N. 1995. Barriers and bridges to learning in a turbulent human ecology. In *Barriers and bridges to the renewal of ecosystems and institutions*, edited by L. H. Gunderson, C. S. Holling, and S. S. Light. New York: Columbia University Press.

Spies, T. A. 2004. Ecological concepts and diversity of old-growth forests. *Journal of Forestry* 102(3):14–20.

Stankey, G. H., and B. Shindler. 1997. *Adaptive management areas: Achieving the promise, avoiding the peril*. General Technical Report PNW-GTR-394. Portland, OR: USDA Forest Service, Pacific Northwest Research Station.

Stankey, G. H., R. N. Clark, and B. T. Bormann. 2005. *Adaptive management of natural resources: Theory, concepts, and management institutions*. General Technical Report PNW-GTR-654. Portland, OR: USDA Forest Service, Pacific Northwest Research Station.

Walters, C. J. 1986. *Adaptive management of renewable resources*. New York: Macmillan.

Chapter 19

Nontimber Economic Values of Old-Growth Forests: What Are They, and How Do We Preserve Them?

JOHN LOOMIS

Two economic facts about our remaining old-growth forests have far-reaching implications: (i) They are now relatively scarce, and (ii) they have high values in competing uses. Old-growth forests may further be characterized as standing reserve, consisting of timber, habitat, biodiversity, carbon sequestration, recreation, and a multitude of other ecosystem goods and services. Although some of the goods and services provided by old-growth forests can be supplied by younger forests, the older forests often provide higher levels of many ecosystem services. But what may become increasingly important is that these forests also provide value not just for the productive aspects of their goods and services but also from their mere existence.

Bearing in mind that the authenticity of "old growth" can be eliminated by clearcutting, we thus face a clear but complex conflict of values within old-growth forests. On the one hand are lumber products valued in markets via readily observed prices; sales of these products, of course, provide tax revenues. On the other hand are the many other ecosystem services provided by old-growth forests that are not traded in markets and are therefore not priced. Most of these ecosystem services of old growth don't generate tax revenues comparable to lumber products, at least not yet.

The absence of market prices on environmental values such as wildlife habitat, watershed protection, or public land recreation contributes to the existing misperception that protecting old-growth forests involves a choice between "the economy and the environment." From the standpoint of economic analysis (as opposed to financial analysis), this is a false dichotomy—the environmental and recreational values of old-growth forests do indeed have an economic value to people because they are scarce and provide enjoyment. For economic analysis is concerned with more than just profits and losses; economic principles apply to the optimal allocation of all scarce resources, whether those resources are traded through markets or not. Economics is often considered a behavioral and social science that seeks to understand the broad array of human values, for many of which society has purposely chosen not to establish markets, or which cannot be efficiently allocated through markets. In fact, the characteristics of many ecosystem services from old-growth forests make it difficult for private firms to capture them as market revenues, but they are no less valuable to society than traditionally marketed products.

This chapter describes these economic values and empirical methods to measure their public worth, and ends with some policy proposals that would allow landowners managing for old-growth forests to receive some of this public value. At present, much of the remaining pressure to harvest old-growth forests comes from the fact that only by harvesting them can private landowners capture their private economic benefits as revenue. The policies I propose would allow private landowners to be rewarded for managing old-growth forests that provide environmental values the public desires.

Types of Nonmarket Benefits of Old-Growth Forests

The nonmarket values of old-growth forests can be grouped into three broad categories of benefits:

- On-site user benefits
- Downstream or off-site user benefits
- Off-site nonuser or passive-use benefits.

The *on-site user benefits* of old-growth forests include recreational uses such as hiking, fishing, and hunting. Often the pristine nature of old-growth watersheds results in excellent fishing experiences, both in terms of catch and quality of surroundings. For some game species, such as Sitka black-tailed deer in southeast Alaska and elk, old-growth forests provide prime habitat that produces trophy animals for hunters. Economists can measure the monetary value of the recreation benefits to the visitors by the additional amount of money visitors would pay to hike, fish, or hunt in old-growth forests as compared to the next best substitute recreational setting.

Old-growth forests also provide economic benefits to people who may never set foot in the forest—*downstream or off-site benefits*. Old-growth forests protect municipal watersheds, providing clear, low-sediment water to downstream towns and cities. High water quality provides a direct aesthetic value in terms of clarity and taste and can reduce the cost of water treatment (Moore and McCarl, 1987; Loomis, 1989); as a prime example, the old-growth forests in the Bull Run watershed outside of Portland, Oregon, provide such high-quality water that only minimal treatment is needed. In addition, old-growth forests can provide a scenic backdrop that increases property values of nearby residences. Old growth helps protect some of the biodiversity that is increasingly being lost due to land conversion.

Perhaps the largest aggregate social benefit is the value that people receive from a set of intangible conditions: (i) knowing that preservation of old-growth forests today provides the opportunity to visit them in the future, (ii) knowing that old-growth forests exist, and (iii) knowing that preservation today provides old-growth forests to future generations. These are known as *passive-use* or *nonuse values* because they represent the enjoyment and satisfaction derived without currently visiting these old-growth forests. These values may be smaller per person than those of a hiker in the forest; however, there are literally millions of people who simultaneously receive these passive-use values.

Although technology can develop substitutes for many of the values of old-growth forests, substitutes do not exist for passive-use values. For example, old-growth lumber is prized for its strength as a building material, but laminates of plantation wood have similar strength properties. Water quality can be enhanced through filtration systems. Carbon can be captured by any forest to varying degrees. However, passive-use values are ends in themselves—they are not means to some other end but are the essence of old growth. Old-growth provides enjoyment directly through thinking about the authenticity of these old forests. Further, it is hard to imagine technology offering a good substitute for old growth for most people. If the forests were actually gone, virtual reality could produce a museumlike replica, but for most people it would be a poor substitute, and the authenticity would be absent.

Passive-use values are administratively recognized as compensable economic values under federal legislation such as the Comprehensive

Environmental Response, Compensation, and Liability Act (Superfund) and the Oil Pollution Act of 1990. These values have been upheld by the U.S. District Court of Appeals (1989) as economic values the U.S. Department of Interior agencies are required to consider in their natural resource damage assessments. In fact, it was the U.S. District Court of Appeals (1989) that coined the term *passive-use values* to describe the three values listed above. Given the legal standing of passive-use values for natural resource damage assessment, it seems appropriate that they be included in economic valuation of old-growth forests. The sum of on-site, off-site, and passive-use values is referred to as total economic value of the *public* goods of old-growth forests (Cordell et al., 2005). This distinguishes them from the value of the *private* goods of old-growth forests, such as lumber.

Techniques to Measure Total Economic Value

Passive-use and total economic value can be estimated using a simulated market or simulated voter referendum approach called the *contingent valuation method (CVM)* (Loomis and Walsh, 1997). This approach asks an individual to identify the maximum amount of money he or she would pay to have protection of a specific amount of a natural resource. Entire books have been written on the CVM (e.g., Mitchell and Carson, 1989). Besides describing CVM, these books evaluate the controversies regarding the accuracy or validity of the monetary amounts that people state they would pay in the surveys. Although CVM has been found to be quite accurate for measuring use values (Carson et al., 1996), until recently CVM surveys of passive-use values have frequently overestimated actual cash willingness to pay by a factor of two or more (Champ et al., 1997). This latter research on response uncertainty calibration has provided a technique to improve the accuracy of passive-use value estimates. However, even earlier applications of CVM have been shown to yield reliable estimates of passive-use and total economic values (Loomis, 1990; Reiling et al., 1990).

Empirical Estimates of Total Economic Values

Early estimates of the total economic value of the public goods aspects of old-growth forests were calculated in the context of protecting the northern spotted owl and began to appear in the 1990s (e.g., Rubin et al., 1991;

Hagen et al., 1992). The Hagen et al. study of the public's total economic value of old-growth forests and spotted owl protection involved a survey of 1,000 U.S. households and achieved a response rate of forty-six percent of deliverable surveys. Their simulated old-growth forest program was drawn from the Thomas et al. (1990) report describing the spotted owl as an indicator species for the well-being of old-growth forests. In their survey, respondents were told of the costs of protecting old-growth forests in the form of higher wood prices and higher taxes. Households were asked to vote yes or no to protection at a particular price that varied across the sample. Hagen et al. (1992) estimated that the average U.S. household would pay $86 annually to protect old-growth forests identified as crucial owl habitat in Oregon and Washington. Given that there are more than 100 million households in the United States, the national benefits of protecting old growth equal several billion dollars, even adjusting for nonresponding households.

More recently, Garber-Yonts et al. (2004) estimated that Oregon households would pay $380 annually to increase old-growth forests from their current five percent to thirty-five percent of the forest stock in the Oregon Coast Range. Comparing this $380 to the average of $86 from the 1992 nationwide study suggests that residents in the region where the old-growth forests exist value old growth nearly four times as highly as other U.S. residents. The $380 for old-growth forests in the study also ranked as one of the highest valued land use designations in the survey, much larger than biodiversity reserves, for example. High values for old-growth forests may reflect the public's implicit recognition that protecting old-growth forests simultaneously protects a broad range of ecosystem service values for wildlife habitat, water quality, and climate regulation. Furthermore, these high estimates may reflect high value placed on the concept and existence of old growth itself.

Another approach to estimating the value of old-growth forests requires estimating the value of individual ecosystem services provided by old-growth forests (Heal, 2000). A comprehensive analysis has yet to be undertaken because not all the ecosystem services have been quantified by ecologists. However, Harmon et al. (1990) studied carbon sequestration in old-growth forests, and their findings suggest that conversion of old-growth forests into younger forests would result in a net release of carbon. From the economics side, Costanza et al. (1997), Morton (1999), and Loomis and Richardson (2001) have estimated the benefits from carbon sequestration and climate regulation at between thirty-five and sixty-five dollars per acre of forest of any type. These benefits have become a focal point in the establishment of carbon markets, which will be discussed later.

Policies for Making Preservation of Old-Growth Forests Viable

The nonuse or passive-use values arising from old-growth forests have many of the features of public goods, as opposed to private goods such as lumber. Specifically, the lumber value of old-growth forests can be captured in a market by a landowner because the supply of old-growth lumber can be withheld if the price is not paid by the buyer. That is, a landowner can exclude someone who does not pay for the lumber from consuming it.

By contrast, landowners cannot exclude people from seeing old-growth forests in the distance, nor can they stop people from enjoying the knowledge that old-growth forests exist and are providing habitat for wildlife species dependent upon them. Because they cannot exclude people from enjoying these benefits, people have little incentive to voluntarily pay for these benefits. This is the same rationale that justifies public funding and provision of national defense—once provided, it is available to all, and no one can be technically or economically excluded from receiving the protection. Thus, the only funding mechanism that can ensure that all the beneficiaries pay is mandatory payment in the form of taxes. In the case of old-growth forests, what must be done to provide landowners with an economic incentive to protect these forests on private land is to transfer some of the off-site and passive-use values the public receives to the landowner. Similarly, to fund public land managers to restore and protect old-growth forests on public lands (e.g., protect it from fire, ensure a continuous supply of old growth) is either to tap some of the public value or at least demonstrate that value to Congress so that Congress will provide the optimal amount of funding for the management of these forests.

Incentives to Private Landowners

At least two policy instruments might be feasible in providing economic incentives to private landowners for retaining existing old-growth forests and literally to grow old forests on their land—an old-growth forest reserve program and a deferential property tax program.

- An old-growth forest reserve program can be patterned after USDA's $2 billion annual Conservation Reserve Program (CRP) and $250 million annual Wetlands Reserve Program for farmers. With an old-growth forest reserve program, landowners would be paid an annual amount per acre to enroll their old-growth forests in this program. In exchange for agreeing to manage their forests as old growth, they would be paid a mutually agreed annual amount. This amount would be determined the way it is for CRP: Landowners would propose to the government the number of acres of what age to enroll and indicate the minimum amount they would want to be paid each year for these acres. As is done in CRP these bids would be compared to an environmental benefits index (USDA Farm Service Agency, 1999) associated with the property to select those parcels that provide the greatest benefits per dollar invested. This ensures that not just the lands with the cheapest bids are enrolled without considering that some tracts of land may provide more environmental benefits due to being close to streams, adjacent to other tracts, etc. Forest and wildlife considerations are already included in this index, so adapting it for old-growth forest should be easy.

- Property tax reduction tied to the age of the forest. At present, some forms of property taxes on the value of timber stands increase with the volume (age) of the trees. This represents a penalty for growing older trees. Instead, the tax rate needs to decrease with the age or size of the tree to provide an incentive to allow the trees to grow older and larger. Furthermore, lower property tax rates can be assessed on old-growth tracts, in a manner similar to deferential property tax rates used in the preservation of agricultural land.

Capturing Old-Growth Forest Values to Fund Public Agencies

Although federal and state agencies already own land that contains old-growth forests, protecting these forests from threats such as catastrophic wildfire (Verner et al., 1992) or insects is expensive. Further, if society wants more old-growth forests, significant management effort may be required to restore existing forests and grow more old growth. At present, federal land management agencies' funding is inadequate for these purposes. In the past, agencies' budgets were positively related to how much timber was offered for harvest (O'Toole, 1988) not how much old growth was preserved. Moving public land management agencies' annual budget appropriations process away from line-item budgeting based on specific resources to integrated natural resource budgeting has been recommended as a way to provide more balanced funding (Shands et al., 1990; Loomis, 1993).

Several avenues seem open for procuring this funding:

- Demonstrating to the U.S. Office of Management and Budget and to Congress the magnitude of ecosystem values of old-growth forests to support budget requests for protecting, managing, and restoring old-growth forests.
- Capturing more of the ecosystem service values that old-growth forests provide to recreation users, downstream water users, etc. An example of the potential for value capture is the European Union's carbon trading market, established in 2005 to allow buying and selling of carbon credits as one way for a country to meet its Kyoto Protocol carbon emission limits. Although the United States has yet to sign the Kyoto Treaty limiting emissions, there is a Chicago Climate Mart that hopes to link to the European Union market to facilitate carbon trading among multinational companies (International Emissions Trading Association, 2006). Using the Clean Development Mechanism in the Kyoto Protocol, forest landowners in other countries have been selling carbon credits to firms in industrialized countries. It does seem possible for U.S. private landowners to do the same thing (see Harmon et al., 1990), and possibly for public land management agencies to receive a monetary credit in their budget for carbon sequestered in old-growth forests to be used to offset emissions by other federal agencies (e.g., U.S. Department of Defense).
- Empirically estimating the incremental value of hiking and backpacking in old-growth forests would justify an add-on to recreation fees for these areas. Under the Federal Lands Recreation Enhancement Act, the agency would be able to keep most of this increment for protecting these forest from threats associated with recreation, such as fire. Much like joint U.S. Fish and Wildlife Service and state fish and game agency administration of hunting on national wildlife refuges (where a separate federal Duck Stamp is required in addition to a state hunting license), an old-growth hunting or hiking stamp could be implemented.

The cost savings to downstream municipalities, hydropower producers, and industrial users of the low-sediment, high-quality water emanating from old-growth forests could be calculated and a portion of that cost savings recouped from these downstream users. Specifically, the improvement in water quality associated with protecting old-growth forest from fire might be calculated and a portion of this cost charged to the downstream beneficiaries. In essence, this is a form of the user-pays approach to funding.

Conclusion

Old-growth forests provide many economic benefits, the majority of which are not reflected in market transactions. Some of these benefits are actually received on site by public land recreation users such as hikers, backpackers, anglers, and hunters. However, substantial benefits are also received downstream by municipal, industrial, and residential water users. The vast majority of benefits are received far from the forests themselves by citizens knowing that protection of old growth today provides them the opportunity to visit old growth in the future, knowing that old growth exists, and knowing that protection of old-growth forests today provides these forests to future generations.

A key question facing society is how these substantial values of old-growth forests can be more explicitly recognized by decision makers. The decision makers on public lands are often federal budget agencies such as the U.S. Office of Management and Budget or congressional appropriations committees. Economists presume that, by monetizing the recreation use values, the downstream cost savings from clean water, and passive-use values, these values can be made more equivalent to the commodity values and tax revenues associated with harvesting the forests—and therefore better decisions can be made. For public land management decisions, it is the public that makes the tradeoffs, bearing both the costs and the benefits of decisions. However, the decision makers on private lands are the landowners themselves. For these private landowners, it is important to make preserving old-growth forests more profitable than cutting them. This can be done by establishing an "old-growth reserve" system similar to the Conservation Reserve Program, removing any disincentives from and interjecting positive incentives into the current tax system, and establishing markets for the provision of old-growth benefits.

LITERATURE CITED

Carson, R., N. Flores, K. Martin, and J. Wright. 1996. Contingent valuation and revealed preferences methodologies: Comparing the estimates for quasi-public goods. *Land Economics* 72(1):80–99.

Champ, P., R. Bishop, T. Brown, and D. McCollum. 1997. Using donation mechanisms to value nonuse benefits from public goods. *Journal of Environmental Economics and Management* 33(2):151–62.

Cordell, H. K., J. C. Bergstrom, and J. M. Bowker. 2005. *The multiple values of wilderness.* State College, PA: Venture Publishing, Inc.

Costanza, R., R. d'Arge, R. de Groot, S. Farber, M. Grasso, B. Hannon, K. Limburg et al. 1997. The value of the world's ecosystem services and natural capital. *Nature* 387:253–60.

Garber-Yonts, B., J. Kerkvliet, and R. Johnson. 2004. Public values for biodiversity conservation policies in the Oregon Coast Range. *Forest Science* 50(5):589–602.

Hagen, D., J. Vincent, and P. Welle. 1992. Benefits of preserving old growth forests and the spotted owl. *Contemporary Economic Policy* 10(1):13–26.

Harmon, M., W. Ferrell, and J. Franklin. 1990. Effects on carbon storage of conversion of old-growth forests to young forests. *Science* 247:699–702.

Heal, G. 2000. *Nature and the marketplace: Capturing the value of ecosystem services.* Washington, D.C.: Island Press.

International Emissions Trading Association. 2006. *Chicago climate mart to try CO_2 link with EU,* April 4. http://www.ieta.org/ieta/www/pages/index.php?IdSitePage=1094.

Loomis, J. 1989. A bioeconomic approach to estimating the economic effects of watershed disturbance on recreational and commercial fisheries. *Journal of Soil and Water Conservation* 44(1):83–7.

Loomis, J. 1990. Comparative reliability of the dichotomous choice and open-ended contingent valuation techniques. *Journal of Environmental Economics and Management* 18(1):78–85.

Loomis, J. 1993. *Integrated public lands management: Principles and applications to national forests, parks, wildlife refuges, and BLM lands.* New York: Columbia University Press.

Loomis, J., and R. Richardson. 2001. Economic values of the U.S. wilderness system. *International Journal of Wilderness* 7(1):31–4.

Loomis, J., and R. Walsh. 1997. *Recreation economic decisions: Comparing benefits and costs.* State College, PA: Venture Publishing.

Mitchell, R., and R. Carson. 1989. *Using surveys to value public goods: The contingent valuation method.* Washington, D.C.: Resources for the Future.

Moore, W., and B. McCarl. 1987. Off-site costs of soil erosion: A case study in the Willamette Valley. *Western Journal of Agricultural and Resource Economics* 12(10):42–9.

Morton, P. 1999. The economic benefits of wilderness: Theory and practice. *University of Denver Law Review* 76(2):465–518.

O'Toole, R. 1988. *Reforming the Forest Service.* Washington, D.C.: Island Press.

Reiling, S., K. Boyle, M. Phillips, and M. Anderson. 1990. Temporal reliability of contingent values. *Land Economics* 66(2):128–34.

Rubin, J., G. Helfand, and J. Loomis. 1991. A benefit–cost analysis of the northern spotted owl. *Journal of Forestry* 89:25–30.

Shands, W. E., V. A. Sample, and D. C. LeMaster. 1990. *National forest planning: Searching for a common vision, vol. 2. Critique of land management planning.* FS-453. Washington, D.C.: USDA Forest Service, Policy Analysis Staff.

Thomas, J. W., E. D. Forsman, J. B. Lint, E. C. Meslow, B. R. Noon, and J. Verner. 1990. *A conservation strategy for the northern spotted owl: A report of the Interagency Scientific Committee to address the conservation of the northern spotted owl.* USDA Forest Service; USDI Bureau of Land Management, Fish and Wildlife Service, and National Park Service, Portland, OR. Washington, D.C.: U.S. Government Printing Office.

USDA Farm Service Agency. 1999 (September). *Environmental benefits index, fact sheet.* Washington, D.C.

U.S. District Court of Appeals. 1989. *State of Ohio v. U.S. Department of Interior. Case No. 86–1575.* District of Columbia. July 14, 1989.

Verner, J., K. S. McKelvey, B. R. Noon, R. J. Gutiérrez, G. I. Gould, Jr., and T. W. Beck, tech. coords. 1992. *The California spotted owl: A technical assessment of its current status.* General Technical Report PSW-GTR-133. Albany, CA: USDA Forest Service, Pacific Southwest Research Station.

Chapter 20

Regional Conservation of Old-Growth Forest in a Changing World: A Global and Temporal Perspective

HAL SALWASSER

With the old-growth controversy of the Pacific Northwest still a part of our very recent history, we continue to run the risk of not seeing the forest *or* the trees with any clarity; of not seeing that they move and change; and that they are redefined both by ecological events and by people, across the entire globe. My aim with this chapter is to provide a perspective on old-growth forests in the context of global forests and change through deep time. For this purpose, I consider *old-growth* forest to be any tree-dominated ecosystem that has passed its early and middle developmental growth stages and is clearly in the later stages of its successional development. The old-growth stage of forest development varies by forest type in age, structure, and species composition. This variation reflects species adaptations to growing conditions that are themselves highly variable across the planet. Distinct areas of old-growth forest may occur as large contiguous areas of similar forest condition or as distinct patches in a landscape mosaic of other forest developmental stages.

For some forest types, such as lodgepole pine in the western United States, old growth is not very old. Fire and insects typically "recycle" lodgepole pine forests on roughly a century time scale. For others, such as giant sequoia in the southern Sierra Nevada mountains of California, old growth is very old indeed; the dominant trees are several thousand years old. An old-growth lodgepole pine forest is also relatively simple in species composition and structure compared to an old-growth forest of what is commonly called mixed conifer. Tropical, temperate, and boreal old forests also vary greatly in where their biomass and biodiversity occur. Tropical old forests tend to contain most of their biomass and species diversity above the soil surface, whereas temperate drier old forests have biomass and diversity both above and beneath the soil surface. The point is that the old-growth stage of forest development does not look or function the same across all forest types.

Old-Growth Changes

Regardless of age, structure, or species composition, all old-growth forests are created and maintained by interactions among their constituent species and among those species, by their physical environment and by natural events such as glacial advances and retreats, fires, wind storms, landslides, and floods. Although a given old-growth forest may look "stable" or "in equilibrium" to a casual observer on an annual or decadal time scale, all forests are highly dynamic on multiple temporal and spatial scales, with both species composition and structure changing over time and space. Any perspective on old-growth forest must be taken with those dynamics in mind.

Likewise, human relationships with one another and with the rest of nature are just as dynamic through space and time as we now understand forests to be, perhaps more so. In the course of just the past forty years, for example, human interactions with old-growth forests in the United States have decidedly changed. When I began my professional career as a wildlife ecologist with the U.S. Forest Service in the late 1970s, old growth was seen as decadent and unproductive forest that should be harvested and replaced by faster growing young trees. It was also at about that time that research on forest ecosystems, including old growth, began to illuminate the unique characteristics and functions of old forests. Yet it would take two decades and much political and social turmoil following the appearance of those new findings for them to lead to changes in forest policies regarding old growth. It would be safe to say that forests and human perceptions about their values will continue to change in the future.

If we could move backward or forward in time and look at the entire land surface of the earth as a giant photograph, by taking a sequence of snapshots we would see that what covers the land changes—little changes over short time periods and huge changes over long time periods. Over the past several million years we would see the comings and goings of around forty cycles of glacial and interglacial periods, rising and falling sea levels, and forests moving and changing not only in location but also in species composition. Over hundreds of millions of years we would see entire continents moving across the surface of the earth, isolating or reassembling their biota in the process. Whatever we might consider old-growth forest to be today, it is ephemeral in the grand sweep of time; nature has never kept it stable or in the same place for all time.

People as a Force of Change

Prior to several million years ago the land surface cover we would see with our snapshots would have been whatever nature absent human beings delivered. But with the emergence of early humans (*Homo erectus*) and their eventual dispersal out of Africa into Europe and Asia around one million years ago, nature ceased being the only driver of change (Williams, 2003). Human influence on forests through use of fire, hunting, and gathering would most likely have been slight and localized at the beginning, then spreading and more pervasive as modern humans (*Homo sapiens*) subsequently moved out of Africa across the European, Asian, and Australian continents an estimated 50,000–70,000 years ago. In some places, they replaced earlier hominids who appear to have used fire extensively, and in other places they were the first humans to show up. Modern humans appear to have arrived in the Americas perhaps as early as 15,000–20,000 years ago, prior to the end of the most recent glacial period (Shreeve, 2006).

At whatever time and place humans and their use of fire and hunting tools arrived, it marked the beginning in those places of a new force of ecosystem change: human action. Some human-caused forest transformations are direct and long lasting—for example, converting entire forest landscapes into farms, as occurred in the U.S. Southeast, Northeast, and Midwest beginning in the 1600s. Others are more ephemeral, such as shifting agriculture in tropical rainforests. Some permanent transformations may also have resulted from human-aided species extinctions, especially on islands or where the native biota did not evolve over long periods with human coexistence. Large mammals and their predators and large flightless birds seem to be especially vulnerable to the effects of human occupancy. These species are either herbivores or eat herbivores to subsist. Their loss likely created cascading changes to ecosystems through changes in food webs whether or not those ecosystems had been transformed directly through fire or early agriculture. A recent example of this in the United States in the Great Lakes states, the Northeast, and some parts of the South is the dramatic effects of white-tailed deer on forest understory in the absence of their natural predators.

The point here is that humans directly transform ecosystems through use of fire or conversion from one ecosystem type to another. They can also set in motion indirect changes that result from altered food webs, such as the deer example above. In this light, given the long occupancy of most forested areas of the world by humans, one must envision forests in periods prior to human occupancy to get a sense of what a truly pristine old-growth forest unaffected by human activity might look like. It is not likely to be the kind of forest the world has seen for many millennia.

Human influences on forests dramatically increased in scope and magnitude following the most recent glacial cycle. Beginning at least 8,000–10,000 years ago, people learned how to replace their nomadic and hunter–gatherer–angler lifestyles with stable communities based on sedentary rather than shifting agriculture. This move appears to have started in the Fertile Crescent of Mesopotamia, the Indus River valley in what is now western India and eastern Pakistan, and parts of eastern China. It appeared at least 3,000–4,000 years ago in several areas of Central and South America (Mann, 2005). It is most likely that early people made their first farms in woodland openings, on floodplains, and on human-enhanced terraces near water but above flood levels. However, as populations and settlements grew, people's needs for land for farms and wood for fuel, farm implements, weapons, and building materials increasingly affected more forest area.

Working with Half the Forest

Why start a perspective on contemporary old-growth forest conservation with this prehistorical context? Because human population growth, expansion, and land transformation have likely resulted in the loss of as much as *half* of the global forest cover that may have existed 8,000 years ago (Williams, 2003). Only a portion of the lost forests would have been old growth, because the combination of natural events and human activities would have created and maintained perpetually dynamic mosaics of different-aged forests across the landscape. Nonetheless, some of the forest at any place and point in time prior to agriculture must have been old-growth or late-successional forest. But about fifty percent of the world's forests prior to postglacial growth of the human enterprise is not forest of any kind anymore. Thus, at best, any attempt to perpetuate old growth as part of a dynamic forest landscape must start with knowing that in many places we are working with only a fraction of what might have been in place had the human enterprise not evolved as it has. It must also start with knowing that humanity is not a mere 5–10 million souls, as is estimated for the preagricultural global population, or the 500 million estimated at the dawn of the industrial era a mere 300 years ago, but nearly 6.5 billion, heading for perhaps 8–10 billion by mid-century.

That is the prehistorical perspective on old forests. What about today and pressures on remaining forests? The most current FAO report (2005) estimates that as much as 17 million acres net of global forests were converted to other land uses annually between 2000 and 2005. This calculation incorporates an estimated 32.5 million acres lost largely to agriculture in tropical forests and 15 million acres added in temperate forests through return of abandoned agricultural lands to a forested condition. Clearly, population growth and agricultural expansion in some developing countries continue to come at the expense of natural forest, while domestic production efficiency combined with global trade in food, wood, and industrial products allows economically affluent countries to expand their forest cover even when consumption of forest products grows.

Forests lost in developing countries tend to be older, more natural forests while forest accretion in developed countries typically involves planted forests of either native or exotic species. The forests being lost in developing countries may have been transformed to some degree by human activities such as shifting agriculture, domestic stock grazing, selective harvest of native biota, or use of fire, but they were most likely more "natural," species-rich forests than the planted forests that are partially offsetting their loss in both tropical and temperate zones. The global story on forest loss and forest gain is that it is not an ecologically balanced exchange.

Got Wood?

Demand for wood products and persistent human pressures on forests are the most significant forces of change in today's forests. But their potential negative impacts on forests can be alleviated. Let's look at international demand and trade first. International trade in wood began several thousand years ago as early cultures sourced wood for buildings and ships from other cultures. It accelerated with the increasing use of wooden ships for war and industrialization in recent centuries. International trade in wood products is now a global enterprise that entails moving products, across at least one international boundary from the time they are a tree to a finished product (fig. 20.1). Much of this trade has historically involved the U.S.–Canada exchange.

The FAO (2005) estimated that 56.5 billion cubic feet of industrial wood was produced and consumed in 2000. To envision that quantity of wood, imagine an American football field covered by a woodpile more than 200 miles high, including the end zones! About one-third of current global wood production comes from forests planted for the purpose of growing trees for wood products. The other two-thirds comes from forests where the trees were not planted by people. Some, but a declining percentage, of these forests are being harvested for the first time—in other words, they are likely to be old growth. Until the middle 1900s, most industrial wood products used in the world came from trees in forests grown by nature alone. By 2050 the world could be obtaining more than seventy-five

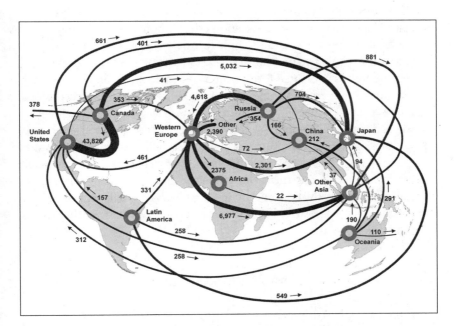

FIGURE 20.1. Major net trade flows of sawn softwood in 2000 (in million cubic feet). Based on figure from Food and Agriculture Organization (FAO) of the United Nations FAOSTAT Trade Flow Data.

percent of its industrial wood products from planted forests covering a relatively small area of land (Victor and Ausubel, 2000).

Whether these new planted forests are composed of exotic or native tree species, they are typically growing on only the most productive soils, in places with affordable access to manufacturing facilities and markets, and in nations with relative political stability and secure land tenure. Unlike in forests not planted by people, intensively managed planted forests involve raising superior seedlings, preparing growing sites, and planting and tending a new tree crop. These activities require capital investments that flow only to places delivering high returns on investment at acceptable risk.

It is now, or soon will be, technologically feasible to deliver wood yields of up to 286 cubic feet per acre over the life of a typical crop rotation on the best tree growing sites in the world, such as temperate forests in the U.S. South; the coastal U.S. Pacific Northwest; and afforested agricultural lands (planted forests) in temperate parts of China, India, Brazil, Chile, South Africa, New Zealand, and Australia. Even given a high projection for global demand for wood products by 2050 at 71 billion cubic feet, producing that much wood on only those acres that can deliver 286 cubic feet per acre would require only 247 million acres, 2.5 percent of the world's current forest land. Using more conservative and perhaps more realistic average yield estimates, growing wood at 143 cubic feet per acre over the crop rotation period requires only five percent of the world's most productive forest.

Future trends in planted forests, intensive silviculture, and manufacturing and marketing efficiency mean the possibility of much reduced global reliance on natural forests as sources of industrial wood. Thus, focusing wood production on planted forests could alleviate the need to harvest trees in natural forests for commodity wood products, allowing many to grow to older ages with little or no expectation for eventual harvest or perhaps only selective harvests. If this turns out to be the case, producing the world's industrial wood products primarily from a relatively small area of planted forests could have a positive impact on forest conservation for other values, including old growth.

Possible Wood Futures

The U.S. Pacific Northwest has some distinct advantages in such a scenario. First it still has a relatively large area of unharvested, old conifer forest that is reserved from harvest on federal lands. The planted forests of the region on private lands are composed of native species that produce solid and engineered wood products along with pulp and paper. Other places that can grow trees fast are doing so with exotic species and may produce only pulp and paper raw materials. Thus, the Pacific Northwest region could become a unique place where significant old forest conservation occurs on the same regional landscape as highly productive forest growing trees for global wood markets.

Should this future of focused wood production coupled with larger areas of forest reserves occur, it would still not solve problems created by other human pressures on global forests. Use of wood for fuel consumed an estimated 64 billion cubic feet of wood in 2000 and typically comes from community forests grown for fuelwood production or from lopping reachable limbs or gathering fallen limbs and branches in natural forests. These latter activities, often accompanied by year-round grazing of domestic livestock and frequent use of fire to stimulate forage plants, continue to severely degrade natural forests in many developing nations. Illegal logging, and human settlements that follow road building in countries with poor law enforcement, also remain significant forces of forest change in developing tropical nations as well as the boreal forests of Russia.

In the United States, the greatest current pressure on forests is land use conversion to residential use, better known as *urban sprawl*. Stein et al. (2005) estimated that up to one million acres of forest per year were converted to residential use in the United States between 1990 and 2000. Some of this loss may be unavoidable due to rising real estate values. But keeping domestic wood products businesses competitive in global markets would appear to increase the likelihood that landowners will continue to invest in forestland for forest and wood uses. Thus, a forest products sector that maintains competitive advantage in world markets based on highly productive *planted* forests, superior quality products, rapid access to major markets, and top customer service, if it reduces harvest pressure on natural forests as described above, would also be a positive force for old-growth forest conservation. This appears to be the case in the U.S. Pacific Northwest, where most wood products companies now rely on supplies from private lands and have aligned their processing technologies and markets in ways that no longer rely on severely constrained supplies of larger diameter trees from federal lands.

Operating in global markets means that companies must compete against products from countries where labor and regulatory costs are a fraction of those in North America. If companies begin to lose competitive advantage, through either increased costs or decreased product value, the region could lose more of its forest management capacity, processing infrastructure, jobs, and economic vitality in rural communities. The world would also burn more fossil fuel, moving wood products into the country with the greatest wood consumption in the world. One counterintuitive result of these conditions: For those older forests in the Pacific Northwest region that require periodic management intervention to reduce vulnerability to climate warming, insects, or fire, losing wood products industry capacity may not be in the best interest of old forest conservation. For other old forests not so prone to drought, insects, or fire, maintaining them in a landscape of other forest uses, even in industrial forest plantations, may help sustain their biological diversity better than nesting isolated old forest reserves in a landscape of residential or agricultural land uses.

Perpetuating Old-Growth Forest in Dynamic Landscapes May Require Management

I opened this perspective with descriptions of old-growth forest variability and change. I then tried to place forests in the context of a world filling with people who augment nature's change agents and—in human time scales—permanently alter the landscape. I described the major transition under way in how the world will likely grow its industrial wood in the future and where it will likely come from. This could be the silver lining in the cloud created by projected human population growth and demand for forest products. If the world does indeed get most of its future wood from intensively managed planted forests, a very large area of forest can be reserved for nonindustrial purposes such as biodiversity conservation, recreation, watersheds, and spiritual retreats. Several authors have suggested this could be as much as fifty percent of the world's currently forested area. But does this mean we just walk away from forest reserves and let nature do its thing? The answer is probably yes for some places and probably no for others. Let me try to explain why it may be no for some forest reserves.

Hundreds of millions of people still live in conditions in which they directly depend on forests for daily food and fuel. Until their quality of life improves, their pressures on forests will not abate. Also, six to eight billion people moving large amounts of goods around the world increases the flow of nonnative plant and animal species into novel environments. Some of these plants or animals will become pioneering invasive species that further transform receiving ecosystems. A growing population could also mean that people will live even more dispersed across the landscape, posing additional challenges for forest managers. Most people do not like wildfire near their abodes or property, and they tend to dislike air pollution caused by fire, whether managed or wild. A management challenge for reserved forests in impoverished countries will be to lift the quality of human life sufficiently to take people out of their daily subsistence mode based on forest resources. Another challenge in places such as wilderness or unroaded public forests in affluent countries may be to bolster ecosystem resilience in the face of natural disturbance processes or learn how to restore surrogates for some of nature's least desired forces of change.

Each challenge will require new approaches to managing forest ecosystems for natural values in dynamic landscapes. For old-growth forest to persist, it must be able to respond with resilience to disturbance and long-term climate change. Among these responses must be the ability to change location from time to time. Perpetuating some older forests will require a combination of conditions within stands that enhance resilience against low to moderate disturbances and redundancy of developmental stages across landscapes so that when major stand-replacing disturbances do occur, there are forests moving into older stages somewhere else to replace the altered stands. Where fires have been suppressed for long periods of time, selective removal of some biomass followed by use of managed fire—perhaps harkening back to prehistorical land use practices—may be needed to restore stand resilience.

Perpetuating regional old-growth forests in a future world of eight billion people will depend on a combination of practices, including (i) focusing future wood production on only the most productive tree-growing sites, thereby relieving harvest pressure on other forests; (ii) acting as a global community to alleviate human pressures on forests by lifting the quality of human life in less-affluent nations; (iii) restoring nature's processes, or surrogates for them, to forests reserved for natural values; and (iv) planning for and sustaining landscapes that contain both younger forests on a developmental path toward old growth and current old growth. The history of human effects on the landscape will not lessen as we move forward, as each of these practices attests. The challenge will be to design management strategies that clearly consider global, regional, and local changes—simultaneously.

LITERATURE CITED

FAO. 2005. *State of the world's forests 2005.* Rome: Food and Agriculture Organization of the United Nations.

Mann, C. C. 2005. *1491: New revelations of the Americas before Columbus*. New York: Knopf.

Shreeve, J. 2006. The greatest journey. *National Geographic* March:62–9.

Stein, S. M., R. E. McRoberts, R. J. Alig, M. D. Nelson, D. M. Theobald, M. Eley, M. Dechter, and M. Carr. 2005. *Forests on the edge: Housing development on America's private forests*. General Technical Report PNW-GTR-636. Portland, OR: USDA Forest Service, Pacific Northwest Research Station.

Victor, D. G., and J. H. Ausubel. 2000. Restoring the forests. *Foreign Affairs* 79(6):127–44.

Williams, M. 2003. *Deforesting the Earth*. Chicago: University of Chicago Press.

Chapter 21

Moving Science and Immovable Values: Clumsy Solutions for Old-Growth Forests

DENISE LACH

After years of science, litigation, and innovative management, what do citizens of the Pacific Northwest (PNW) "know" about old-growth forests? We "know" that the ancient trees are almost gone from public lands after years of clearcutting. That big, old trees are critical habitat for threatened and endangered species. And that old-growth forests happen automatically if you just leave trees alone. Whether any of this is factual in a scientific sense is not of much concern to most citizens. We believe that more old growth is better than less old growth. Better for what? But in the rough-and-tumble world of PNW forest policy in the last decades of the twentieth century, we didn't have the time and, in many cases, the knowledge to answer that question in any sophisticated way.

Science continues to teach us many things about old-growth forest ecosystems. Its methods have brought us reliable—although not infallible—knowledge, describing what happens in ecosystems and sometimes even explaining why it happens. Science does this through several assumptions. The first, and perhaps the most basic, is that the world tends to operate in a linear fashion that can be understood as cause and effect. Science also assumes that the whole—human, automobile, or old-growth forest—can be divided into parts for study, and those parts can be studied independent of the whole. And, finally, it assumes that the sum of the small parts is equal to the whole. If we know how each of the small parts works, we should be able to put them all back together and explain how the whole thing works (see, e.g., chapters 3, 5).

The more scientists learn about how ecosystems work, however, the more they struggle with trying to apply traditional science to a phenomenon that may break some or all of these assumptions. In forested ecosystems, for example, scientists find mutually implicated variables; climate change, for example, increases the length of the fire season and fuel buildup, both of which increase the severity of forest fires, which increases the fuel buildup (high severity fire often leads to flammable, young homogeneous vegetation), etc. It is difficult for scientists to separate relative effects in this situation.

As described in other chapters, new approaches to forest science are describing dynamical systems of elements and subsystems that are defined primarily by their relationships. In turn, these relationships are based on inclusion in hierarchies, scales, and functions that fluctuate depending on the context. Findings from emerging science often appear to contradict what we thought we knew about old-growth forests. And there are scientists who disagree with these new ideas. How should a nonscientist—someone who has to make a policy or life decision—deal with the uncertainties inherent to understanding old-growth forests?

Certainty as a False Idol

Where did we get the idea that science could provide certain knowledge about the world? Traditional models of science—what Thomas Kuhn (1970) called *normal science*—consist of solving puzzles within a well-established "paradigm." Each paradigm includes facts, assumptions, and expectations of how the world works; it generally doesn't allow for other ways of knowing the world—or even questions that can't be answered within the paradigm. The "scientific method" as we know it pursues questions that rest within an existing body of knowledge—advancing our knowledge one experiment at a time. This is the model of science found in most textbooks. By the time we finish our education—through high school or graduate school—most of us have this view of science as a stable, accurate, unbiased collection of facts.

The science of systems is relatively new. Over the past fifty years, scientists have been developing ways to study how systems work as systems—not just how the parts of the systems work. This is particularly difficult, because not just what we know but probably how we think about the world is based on ideas about cause-and-effect relationships. We do know that system relationships include those in which cause and effect are distant from each other in time and space. And in that intervening time and space, we don't really know what's happening to the phenomena in which we're interested. How can we hope to understand all the countless variables involved in an old-growth forest system?

A central question as citizens and/or policymakers is how do we deal with the fundamental uncertainties of science? Two suggestions emerge: (i) reduce our expectations of science, and (ii) develop a better taxonomy of scientific uncertainty that can guide decisions.

Living with Uncertainty

In recent years, it has become common for defenders of the status quo to argue that scientific information pertinent to an environmental claim is uncertain, unreliable, and fundamentally unproven. This lack of proof is used to deny demands for action. But the idea that science could provide proof upon which to base policy is a misunderstanding (or misrepresentation) of science, and therefore of the role that science can play in policy. In all but the most trivial cases, science does not produce logically indisputable proofs about the natural world. At best it produces a robust consensus based on a process of inquiry that allows for continued scrutiny, reexamination, and revision.

What if we thought about science as an intellectual consensus about a specific topic by a group of relevant experts? We would see scientific consensus arising when empirical evidence from multiple sources and tested methods starts to coincide. Of course, there would be some agreement about how problems are framed, including what implicit and explicit assumptions each scientist brings to the research, and also whether there is "theoretical integrity" with existing beliefs and commitments. Actually, this is pretty much how science works in the day-to-day life of scientists; most of us do not stumble upon "facts" or "truth" that we can publish for the world to use. Instead, we spend our time trying to solidify or extend what is already known about the world.

This "reasonable expectations" model of science does not suggest that the goal of science is unanimity—there will always be scientists who dissent with the common understanding. This is good for science; it's how new approaches and ideas arise. It's probably better to think about dissenting scientists as "outliers." In statistics, an *outlier* is an observation that occurs a long way from other values; it's not necessarily a wrong observation, just one apart from the norm. There will also always be *anomalous* results—observations that don't meet our theoretical or empirical expectations. So how do we respond to the presence of dissenters, outliers, and imperfect data in science?

A preliminary taxonomy of uncertainty has been proposed that assesses the general level of consensus about scientific ideas:

- *Science is not generally accepted,* with active scientific debate among scientists. An ecological example is the management of forests after a fire. Some scientists claim that removal of large wood is needed to reduce risk of high-severity fire, but others disagree and claim that post-fire logging can increase fire hazard (Thompson et al., 2007).
- *Science is mostly accepted by scientists,* with perhaps some outliers. A good example is global warming. Although there is a large body of scientists who agree that the planet is warming, a few dissenters argue that changes are within the natural range of variability.
- *Science is contested by outside parties,* with nonscientists such as agency representatives, lawmakers, interest groups, and informed citizens challenging scientific methods, results, and interpretations (see, e.g., chapter 14). The emerging ecosystem-based understanding of old-growth forests (e.g., the importance of a wide variety of habitat elements or landscape patterns) may be an example of this type of uncertainty. Although many scientists are finding the shift to an ecosystems approach valuable, nonscientists aren't sure how to use the ecosystem-based results to further interests such as protecting endangered species or salvaging burned logs (e.g., Chase, 1995).

When there is active scientific debate, the appropriate response is more research. Although it's not a guarantee, additional knowledge has the potential to increase intellectual consensus. In this case, science is not quite ready for prime time—we shouldn't expect to use it to make personal or policy decisions. When science is mostly accepted by scientists with a few dissenting outliers, additional scientific research is unlikely to decrease uncertainty. In these cases, the disagreements are likely to lie in the way problems are framed, what counts as evidence, and how data are interpreted. As problems are reframed and new approaches developed, it may happen that the scientific consensus is shown to be wrong. The new information enters the ongoing process of consensus development.

For those issues in which the science is contested by outside parties, the issues at stake are almost certainly not technical. Instead, they are moral, political, religious, economic, or aesthetic. More technical research will typically not resolve these disputes; people have chosen their positions for nonscientific reasons, and more information will not convince them they are wrong. Decisions on such issues will have to be made through public processes that draw upon multiple points of view, including scientific ones, but are highly unlikely ever to rest simply on what science says about the problem. Of course, problems in this third category may also intersect with the other categories; nonscientists may harden their opinions around science that is not resolved (e.g., climate change), creating many layers of real and perceived uncertainty.

Many of our thorniest resource issues, including management of old-growth forests, may fit in these last two categories of uncertainty, those for which additional research is unlikely to resolve the problems. For example, existing spotted owl habitat in some drier forests is beginning to be understood by scientists as an unsustainable artifact of fire suppression. If we want to have owl habitat (i.e., dense older forests) in some parts of these landscapes, we may have to intentionally thin out understory trees. Removal of smaller trees through thinning may reduce owl habitat at the site level but reduce the risk of owl habitat loss at the landscape level. In the short term, we may end up with diminished spotted owl habitat at some sites and less "natural" forest. And this is all "new science"—yet without intellectual consensus. How should decision makers use this information as they make social and political decisions?

In the past, we looked for "elegant" solutions to natural resource issues: a technological, political/legal, or scientific "fix." We assumed that the best solution would be simple and obvious, although maybe difficult to implement due to hardened social systems and infrastructure. Elegant solutions to systems problems, however, tend to create more problems even as they solve the presenting problem.

The document *Forest Ecosystem Management: An Ecological, Economic, and Social Assessment Report of the Forest Ecosystem Management Assessment Team* (FEMAT 1993), for example, used traditional decision processes to create a suite of policy options to manage federal old-growth forests. Although FEMAT appeared to "solve" the problem of threatened spotted owls by creating a reserve of old-growth habitat, it also created new problems in other system elements. These unintended consequences, discussed throughout this book, include a plan based on "reserves," which created a mentality of no action when in fact fast action is needed to maintain some old-growth types. In addition, reducing the supply of big trees helped speed the development of substitute products (e.g., glulam beams) (see chapter 7) and reduced the demand for and price of large-diameter trees, making it less likely that private owners would grow big trees on their land; a loss of mill capacity and employment made it more likely that there will be little industrial capacity available to handle restoration activities. We were convinced that protecting habitat for the owl would do the job, but socio-economic, legal, and even ecosystem forces (e.g., invasion of the barred owl into spotted owl habitat) outside the control of the land managers have limited the effectiveness of the habitat solution. Although most people and organizations have largely accepted the Northwest Forest Plan (NWFP) as the rule of the land, it does not necessarily ensure sufficient owl habitat or even stop the logging of old-growth trees.

The search for elegant solutions is an artifact of linear and "rational" thinking. It assumes that a problem can be isolated and an optimum solution created. But when we are trying to resolve system problems—or problems embedded in larger systems—rational processes are rarely totally satisfactory and often generate new problems that we didn't previously need to worry about.

Worldviews and Decision Making

We have difficulty making social and political decisions because people bring different experiences, knowledge, and expectations to the table. One powerful tool for thinking about the underlying differences is a framework that has been tested in many countries and appears to be a robust characterization about worldviews and resulting values (Douglas, 1982).

As a social species, human behavior is determined by how we relate to groups: How connected are we to strong groups that bind our decisions? And how constrained are we by those ties to the groups? Figure 21.1 describes how these two dimensions generate four basic and stable forms of worldviews. The horizontal axis describes our preferences to work individually or in groups; the vertical axis relates to how critical we believe external restrictions are to social relationships.

Individualists stress the autonomy of individuals, especially their freedom to bargain with others. They care more for the bottom line than for relationships with people who come together to achieve results. Most of

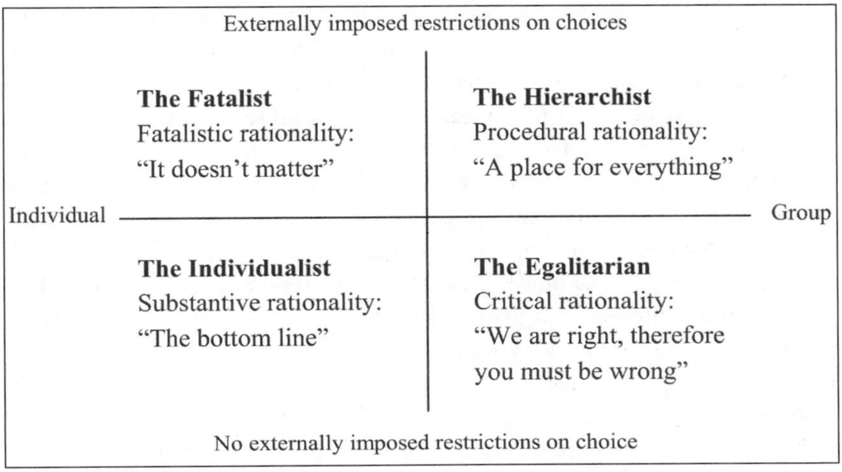

FIGURE 21.1. Dimensions of social relationships. (adapted from Douglas, 1982)

our organizations, however, are made up of orderly and ranked relation-ships, which are most comfortable for *hierarchists*. Managing relationships requires regulations and tracking who does what. *Egalitarians* stress the values of cooperative and volunteer relationships, believing that rules will emerge as needed out of specific situations. Finally, there are marginalized members—*fatalists*—who feel that they have no capacity or ability to influ-ence events. Anything that happens is simply to be enjoyed or endured, never achieved.

Clumsy Decision Making about Old-Growth Forests

Taking what we know about decision making under varying conditions of uncertainty, systems science, and different worldviews, how can we shape decision processes for thinking about the future of old-growth forests? Each idea suggest that it may be most fruitful to design processes that are multiple iterations of problem framing and solution seeking—starting, stopping, circling back—to create solutions that appeal to the different val-ues brought by constituents.

Various factors would seem to promote this kind of "clumsy" solution seeking. The first would be the conviction among participants that they have no viable choice other than to accommodate principles (e.g., profit seeking, rule making, consensus) that at first glance seem anathema. Individualists would have to give up the simplicity of markets, egalitarians accept the existence of unfair differences, and hierarchists cede some amount of power. Even as doctrinal sacrifices are made, participants must see progress, avoid big losses, and not see other interests being unduly privileged. Let's see how suited old-growth forest issues might be to clumsy solution making.

Accumulating and growing problems: Mother Nature can be blamed for catastrophes—forest fires, droughts, landslides. However, when problems occur so frequently that they appear endemic, they begin to be interpreted as manageable human failures. Since the implementation of the NWFP in 1994, issues related to old-growth forests have slipped below the radar for most citizens. However, logging in previously unroaded areas, fire salvage logging of old-growth forests, the Bureau of Land Management plan revision process, and a new U.S. Fish and Wildlife Service recovery plan for the northern spotted owl are bringing PNW old-growth forests back into the headlines.

More to lose by inaction: Public consciousness about old-growth forests rose dramatically with the listing of northern spotted owls and marbled murrelets. Since then, changes in the timber industry (e.g., consolidation, technology, global investments) have created a sector with fewer large *and* small companies, increased reliance on private forests, and a shift to small-diameter logs. Federal and state agencies have been implementing the NWFP for more than a decade, with a large reduction in the harvest of old-growth forests and its attendant income for rural communities. Some communities prosper by moving into new sectors or recruiting new companies, while others continue to suffer the loss of public land harvest. Environmentalists focus their protests and challenges on local harvest decisions but continue to monitor decisions made at higher levels. At this point, these oft-conflicting interests may be convinced that they have nothing left to relinquish in any further compromise, therefore making radical solutions more appealing.

Experience with settlement: In place since 1994, the NWFP was designed to develop cooperative planning, improved decision making, and coordinated implementation with federal agencies and state, tribal, and local governments as "they seek to implement management strategies on forest lands" (USDA and USDI, 1994). Forest management agencies reached beyond their usual constituencies to include recreational users and environmentalists, as well as fish and wildlife agencies and groups. Over time, participants came to know each other through implementing the NWFP, especially at the local level, and a range of values gained legitimacy in decision processes.

Leadership willing to take risks: The kind of leadership needed to craft clumsy solutions is different from that usually found in bureaucracies. What characterizes "clumsy" leaders is not their roles, positions, or access to resources but instead their willingness to take risks that appear to challenge traditional ways of doing business. The motivations for risk-taking are difficult to generalize, because they arise out of particular contexts. Leaders recognize the accumulation of previous failures and understand the unsupportable costs of continued stalemate. Further, they have participated in effective collaborative efforts. These leaders realize that creativity is required for finding viable solutions.

What is not clear at this point is whether federal and state bureaucracies are willing to continue the search for innovative solutions to the wicked problems facing old-growth forests or if private industry is willing to support any changes in the laws, regulations, or accepted practices. Many obstacles stand in the way, including different objectives for local and federal policies, different goals for different landowners, and lack of convening structures that aren't affected adversely by the Federal Advisory Committee Act, which ensures that bureaucratic rules are applied to the smallest gathering at which federal agency personnel are attending. The concept of "clumsy" solutions may itself stand in the way for many professionals who are criticized enough for decisions; they are trained in traditional decision processes, have difficulty integrating non-traditional information into formalized (and legalized) decision making, and ultimately have to justify their decisions to elected officials and the electorate.

Where Are We Now?

Decision makers in the PNW—public agencies, private companies, non-governmental organizations, and individuals—have lived through the "old-growth wars" of the late twentieth century. They attempted elegant solutions for wicked problems—sustained yield management, cessation of harvest on public lands, helicopter logging, litigation—with only limited success. PNW decision makers continue to learn more about the function of old growth as scientists begin to ask new questions about dynamical relations in forest ecosystems. Integrating new ecological information into institutions, practices, and regulations, however, is likely to be difficult because it challenges existing—and tightly negotiated—social, economic, and political relationships.

Traditional roles for science in policymaking have been to provide a broad information base for decisions or to provide information that is put to use for specific projects (e.g., Romsdahl, 2005). Although these roles may still be appropriate for some individuals or some components of clumsy decisions, scientists may be asked to take on different roles as participants craft solutions that recognize and intentionally manage for unintended consequences of policy choices. Scientists can bring their knowledge of complex systems, emerging properties, and temporal/spatial scale effects to discussions of long-term impacts of decisions. They can also help decision makers understand the impact of policy choices on different components of the system. If the choice is to create reserves to protect spotted owl habitat, for example, social scientists can help explore the consequences to communities and industry sectors, while ecosystem scientists help explore the long-term consequences of reserves on a range of system components. Although they won't be able to provide specific answers to the questions posed by policymakers, they can bring their knowledge and habits of thinking to the quest for acceptable policies.

Considering the four conditions described above, the wicked problems of old-growth forest management have been "tamed" for the past decade through the NWFP and are not generally seen as critical issues by most people; however, the challenge has not been neatly resolved and will be reappearing in differing guises in the near future. PNW decision makers probably believe they have yet more to lose by further inaction and have plenty of experience with negotiated settlements. What is not clear, however, is whether leaders at the local-, state-, or federal-level or private landowners with clout enough to influence policy decisions, are willing to take new risks related to old-growth forests, especially if no pressure is forthcoming from constituents, neighbors, or other stakeholders. Old-growth forest management would appear to be a prime candidate for clumsy solutions, with the key question being whether the right kind of leadership exists in either the public or private realm to build such solutions.

LITERATURE CITED

Chase, A. 1995. *In a dark wood: The fight over forests and the rising tyranny of ecology.* Boston: Houghton-Mifflin.

Douglas, M. 1982. *Essays in the sociology of perception.* London: Routledge & Kegan Paul.

FEMAT (Forest Ecosystem Management Assessment Team). 1993. *Forest ecosystem management: An ecological, economic, and social assessment-report of the Forest Eco-*

system Management Assessment Team. 1993–793–071. Washington, D.C.: U.S. Government Printing Office.

Kuhn, T. 1970. *The structure of scientific revolutions.* Chicago: University of Chicago Press.

Romsdahl, R. 2005. When do environmental decision makers use social science? In *Decision making for the environment: Social and behavioral science research priorities,* edited by G. D. Brewer and P. C. Stern. Washington, D.C.: National Academy of Sciences.

Thompson, J. R., T. A. Spies, and L. M. Ganio. 2007. Reburn severity in managed and unmanaged vegetation in a large wildfire. *Proceedings of the National Academy of Sciences* 104(25):10743–48.

USDA and USDI (U.S. Department of Agriculture, Forest Service; U.S. Department of the Interior, Bureau of Land Management). 1994. *Record of decision for amendments to Forest Service and Bureau of Land Management planning documents within the range of the northern spotted owl.* (Place of publication unknown).

PART V

Managing an Icon

In forestry, change really occurs on the sites where the trees are. However, given the ever-changing dynamics of forests and the legacies of past human actions such as logging and fire suppression, there is no simple recipe for how humans can maintain and restore old growth across the landscape. This section considers how evolving science and social knowledge might be used to change the course of forest development. It considers where management can and cannot contribute, and it examines what roles private forest owners play in old-growth conservation.

Franklin, the founder of old-growth ecosystem research, illustrates how reserves and restoration can be used in moist and dry old-growth forests to help us save what we have and develop it where we want more. He argues that active management is needed in many cases, especially for dry old-growth forest types, and discusses how climate change might affect these forests. Tappeiner, a leading silviculturalist in the region, reveals how young managed forests can be manipulated to help increase their structural and species diversity and speed the development of old-forest characteristics. He points out that restoration activities in these forests must begin with a good understanding of their condition and that the devil is in the details: Shrubs, tree sizes and spatial arrangements, and the forest floor all matter when one is attempting to manage for ecological diversity and old growth. Oliver, the author of the leading textbook on forest stand dynamics, puts old-growth management into a landscape and global context. He makes a case for seeing old growth as part of a landscape "triad" of intensively managed plantations for wood production, protected forest reserves for old growth, and integrated management areas where both commodity and noncommodity values are produced. Finally, von Hagen, a prominent regional conservationist, argues that we need to think in new ways about how we manage the forests of the Pacific Northwest. She suggests we do not need to choose between fiber production and public values: Development of new markets for nontimber forest products and services can enhance forest diversity and provide economically viable management strategies for private landowners seeking to contribute to the biodiversity typically associated with the iconic old-growth forests.

Chapter 22

Conserving Old-Growth Forests and Attributes: Reservation, Restoration, and Resilience

JERRY F. FRANKLIN

Controversy has swirled around old-growth forest conservation for more than three decades. The controversy is not surprising given the diversity of social values associated with old-growth forests and their ecological variability within and between forest regions. Our expanding scientific understanding of old-growth forests and the ever-changing social and environmental context further complicate the dialogue. Managers and stakeholders struggle to stay abreast of this complex and dynamic topic.

Policies for conserving existing old-growth forests in moister areas of the Pacific Northwest have coalesced primarily around their reservation from timber harvest and protection from wildfire, as reflected in the Northwest Forest Plan.

Appropriate conservation measures for old-growth forests on sites historically characterized by low-severity wildfire are still being debated. Many managers and stakeholders advocate aggressive active management to move forests toward more historic conditions, thereby reducing the potential for uncharacteristic high-severity fires. Other stakeholders oppose such programs because of concerns about allowing significant logging and potential negative impacts on wildlife habitat. Meanwhile, high-severity fires and insect epidemics have eliminated thousands of acres of historically low-severity old-growth forests.

247

Silvicultural activities to accelerate development of structural complexity in young stands within Late Successional Reserves are also controversial. Most Late Successional Reserves include significant areas of twenty- to sixty-year-old plantations, and such activities are allowed under the Northwest Forest Plan. Many scientists, managers, and stakeholders advocate thinning the plantations to speed development of structural complexity (e.g., Pacific Northwest Research Station, 2002); others argue that plantations will develop the desired complexity naturally, and fear management will result in undesirable ecological impacts.

In this chapter, I propose that developmental patterns of natural stands provide guides to appropriate old-growth management. Northwestern old-growth forests are very diverse, however, largely as the result of variability in site conditions and disturbance regimes. Forests growing on moister sites characterized by high-severity, stand-replacement disturbances contrast greatly with forests on drier sites characterized by low-severity fire disturbance regimes, for example. I focus on these contrasting old-growth conditions because they illustrate fundamentally different historic patterns of stand structure and development and, consequently, different management approaches, even while recognizing that these represent the end points of site and disturbance gradients that exist in the Pacific Northwest. High-severity sites dominate west of the Cascade Range and at higher elevations east of the Cascade Range. Low- and mixed-severity sites characterize low to middle elevations east of the Cascade Range and hotter, drier sites in the Klamath-Siskiyou Mountains and southern Cascade Range (Franklin and Dyrness, 1988; Sensenig, 2002).

The term *restoration* is used for silvicultural activities (including prescribed burning) intended to accelerate development of desired ecological complexity in young stands or to recreate sustainable structural and compositional conditions on sites historically subject to frequent, low-severity wildfire.

Restoring Structural Complexity to Forests on Moist Sites

Old-growth forests west of the Cascade Range are relatively well understood (see citations in Franklin et al., 2002, as a starting point). These forests are dominated by massive stems of Douglas-fir and such shade-tolerant associates as western hemlock and western redcedar. Individual structures include live trees, snags, and logs of varied species, sizes, and condition

FIGURE 22.1. Typical westside old-growth Douglas-fir forest illustrating the structural complexity characteristic of such stands. (Photo: J. Franklin)

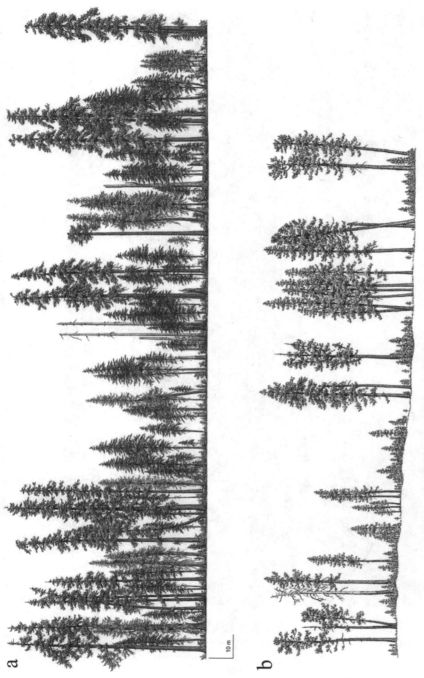

FIGURE 22.2. Cross-sections of old-growth conifer forests illustrating the horizontal and vertical heterogeneity and fine-scale structural mosaic characteristic of such stands: (a) moist westside forest of Douglas-fir, western redcedar, and western hemlock at Cedar Flats Research Natural Area, Washington (maximum tree height is about 270 feet, and cross-section is 890 feet long) and (b) eastside ponderosa pine forest growing on site historically characterized by low-severity fire regime at Bluejay Springs Research Natural Area, Oregon (maximum tree height is about 200 feet, and cross-section is 660 feet long). (Drawings courtesy of Robert Van Pelt.)

(decay) (fig. 22.1). *Canopies*—branches with green foliage—extend from ground level to the tallest tree tops. However, leaves are concentrated in the lower third of the canopy profile rather than at the top—that is, canopies are "bottom loaded"—and shade-tolerant species provide the majority of the foliage. Douglas-firs often have deep crowns because of adventitious or *epicormic* (branches that develop from dormant buds along stems and trunk) branch development.

The complex structure of old growth requires several centuries to develop following a stand replacement wildfire (Franklin et al., 2002). Fires leave behind immense legacies of snags and logs. Trees regenerate, and a forest eventually regains dominance of the site. Young forests experience an extended period of exponential growth following forest canopy closure; smaller trees die from intense inter-tree competition in the dense stands. As stands mature, mortality agents shift toward bark beetles, root diseases, and windthrow, which kill dominant trees, producing canopy gaps and opportunities for shade-tolerant trees to grow into the overstory. Ultimately, the old-growth forest consists of a fine-scale structural mosaic (Franklin and Van Pelt, 2004) (fig. 22.2a).

Much of our information on old-growth forests comes from forests established following major wildfire(s) about 1500 A.D. More than half of the old growth existing when Europeans settled the Pacific Northwest originated at that time. Hence, the "class" of 1500 A.D. strongly influences perspectives on old-growth structure and development, although other old forests exist that originated from wildfires between 1600 and 1800. Many features of the 1500 stands are predictable in any moist Pacific Northwest forest allowed a long period of uninterrupted development. However, many of the 500-year-old stands also display unexpected features, such as a wide age range (e.g., Franklin and Hemstrom, 1981) and high initial growth rates (Tappeiner et al., 1997; Poage and Tappeiner, 2002) in dominant Douglas-firs.

There is strong empirical and theoretical evidence that old-growth development can follow multiple pathways (see chapter 3). Large wildfires are highly variable in severity and include significant areas of low- to moderate- as well as high-severity burn. Hence, there is wide variability in types and levels of live and dead tree legacies, seed sources, and competing vegetation levels within burns. Site productivity also affects rates of structural development. Hence, great diversity exists in patterns of structural development even within areas burned by a single wildfire, let alone in different burns or in areas subjected to multiple burns.

Substantial differences in patterns of Douglas-fir establishment among different cohorts of mature and old forest illustrate the diversity possible in stand development pathways. For example, many Douglas-fir stands, such as those established around 1700, 1845, and 1902, display a relatively narrow age range (e.g., forty to sixty years) rather than the broad age range found in many 1500 stands (Franklin and Hemstrom, 1981).

Some Questions and Speculations About Old-Growth Forests on Moist Sites

Our perspectives on old-growth forest development may be excessively influenced by the 1500 ± stands, which were and still are the most extensive old forests in the region. These stands may have resulted from multiple disturbances occurring over several decades rather than from a single massive fire event, as is the case with many younger, naturally regenerated forests. The scarcity of live tree legacies from preceding stands is possible evidence that the 1500 ± stands originated following repeated wildfires. Lack of seed source could also contribute to the extended period of Douglas-fir establishment (wide age range); either the immense scale of the 1500 ± disturbance event(s) or reburns could have created a scarcity of tree seed source.

Will natural Douglas-fir stands currently 100–300 years in age develop into stands comparable to the 1500 ± stands in 200–400 years? One proposition is that the younger stands will differ because they have higher densities of Douglas-fir and, therefore, slower early growth rates than the 1500 stands (Tappeiner et al., 1997; Poage and Tappeiner, 2002). Hence, the younger stands may have smaller dominant Douglas-firs than the 500-year-old forests—that is, they may become "skinny" old growth.

Concerns about early growth rates of Douglas-fir may be a distraction, however, because there are questions about which structural features are most critical to old-growth forest function. Does the absolute size of Douglas-fir dominants matter most in terms of the functionality of these forests, such as in providing habitat for biodiversity? Or, perhaps, are other features, such as decadence, epicormic branch systems, and heartwood, at least as important as absolute tree size? Perhaps having more numerous old Douglas-firs of smaller size, rather than fewer but larger old Douglas-firs, is better for some ecosystem and habitat functions.

Such questions have practical implications given the current emphasis on thinning young forests to accelerate development of structural complexity. Clearly, there is still much to learn about westside Douglas-fir forests and their development. For example, we know relatively little about structural conditions in stands originating following other wildfires, for example, around 1700, 1800, 1845, and 1902, or about the major shifts in developmental processes in mature (80–200–year-old) stands (Franklin et al., 2002). Each cohort could be—and probably is—highly idiosyncratic in development. And, as noted, variations in stand development are associated with other moist-site forest types, such as the coastal western hemlock–Sitka spruce and montane Pacific silver fir–western hemlock forests, that differ at least qualitatively from Douglas-fir stands.

Accelerating Development of Structure in Young Forests on Moist Sites

Programs to accelerate structural development in young forests can be developed from our substantial knowledge of natural stand structures and development processes. Structural attributes that we can influence include development of large-diameter trees (and eventual sources of large snags and logs) and epicormic branch systems, shade-tolerant species, and spatial heterogeneity. My personal view of moist-site "restoration" is that we can accelerate development of some of the structural complexity and compositional diversity found in older forests. However, recreating old-growth ecosystems is not a credible goal given our still-limited scientific understanding of these ecosystems as well an unsubstitutable requirement for time in the development of some old-growth attributes.

How can we judge the effectiveness of management activities? One objective way would be to assess the degree to which "restored" forest ecosystems fulfill the goals that we have set for them. These may be functional goals, such as in providing for biological diversity. Social goals may also be inspirational rather than ecological. Two caveats here: First, the concept of old forests as places free of human influence is increasingly difficult to defend and, second, creating forest conditions that are both functional and resilient in the face of climate change is increasingly important.

I cannot begin to review the large body of existing information that can inform silvicultural programs to accelerate development of structural complexity in westside plantations. However, I consider the following points important.

Thinning is the most important silvicultural activity—useful to stimulate development of larger trees, epicormic branches, and shade-tolerant tree associates. However, thinning activities to accelerate development of structural complexity will often contrast with the uniform thinning from below usually used to stimulate wood production. Restoration thinning is likely to be spatially variable in intensity, including some areas unthinned *(skips)* and others heavily thinned *(gaps)*. Removal of some codominants and dominants is likely to be necessary to sustain and nurture elements of the understory, including desirable trees in subdominant canopy positions.

Silvicultural activities to enhance complexity can build upon existing diversity within stands. Existing snags, logs, and concentrations of coarse woody debris (pieces of wood larger than about four inches in diameter) can be conserved along with live trees that have or may develop special habitat value, such as stems with structural defects or decay. Potential contributions of trees in lower canopy positions to development of stand complexity need to be considered, including overtopped hardwoods and individuals and patches of high-vigor, shade-tolerant conifers. Similarly, incipient gaps and other areas with well-developed understories are potential foci for heavier thinning.

A wide variety of additional activities and spatial patterns may be appropriate. Examples include killing live trees to generate snags or down boles, or both; wounding or infecting trees to stimulate decay; thinning to release selected dominant trees to stimulate their growth and the development of epicormic branch systems; and planting of desired shrub or tree species, perhaps using gaps opened by overstory thinning.

I offer two caveats in closing this section on moist-site restoration activities:

1) Highly targeted silvicultural activities to create ecological complexity probably require the silviculturalist to be more intimately familiar with the candidate stand than timber production–oriented thinning does, where objectives are usually simpler and better defined. The personal familiarity allows the silviculturalist to effectively build upon the diversity and complexity already present in the stand.

2) The practice of manipulating young forests to stimulate development of structural and biological diversity is in its infancy. Hence, broad-scale application of generic silvicultural prescriptions is inappropriate; this limits opportunities to learn and fails to respect the diversity of forest conditions and societal goals. For example, there are strong regional differences in westside forests in the aggressiveness of western hemlock and local site-based differences in potential for aggressive responses by clonal understory plants, such as

salal and bracken fern. Hence, it is appropriate to vary restoration approaches at multiple scales—stand, landscape, and region.

Restoring Old-Growth Forests on Dry Forest Sites

Stands with old-growth trees on dry sites characterized historically by low-severity fire typically require active management if both the trees and the stands are to be sustained. Furthermore, the old trees themselves and landscape in which the stands are embedded are primary foci in restoring sustainable conditions.

What Do We Know about Dry-Site Old-Growth Forests?

Old-growth forests on dry forest sites that dominate at low elevations east of the Cascade crest and in parts of southwestern Oregon differ greatly from forests on moist sites. These are forests that were historically dominated by ponderosa pine and subject to frequent, low-severity wildfire. On some of these sites, species such as Douglas-fir or grand fir replace ponderosa pine in the absence of fire (Franklin and Dyrness, 1988).

Historically, ponderosa pine and dry mixed-conifer habitat types were dominated by well-distributed populations of large live and dead pine trees with occasional patches of pine reproduction (seedlings, saplings, and poles) (fig. 22.3). Most basal area was in large-diameter (greater than twenty inches diameter at breast height) trees (ten to twenty per acre). Stands were fine-scale, low-contrast structural mosaics (Franklin and Van Pelt, 2004) (fig. 22.2b). Frequent, low-severity wildfires maintained this open structure and dominance of large, old pine trees by retarding tree regeneration and preferentially killing less-fire-resistant species, such as grand fir.

Human activities have drastically changed most old-growth stands on dry sites during the past century (Franklin and Agee, 2003). Changes include dramatic increases in total stand densities, major compositional shifts to shade-tolerant trees, and large reductions in old-pine populations (fig. 22.4). Factors responsible include fire suppression; grazing; logging (especially selective removal of large trees); and creation of dense, even-aged stands by planting.

Important consequences of these changes include (i) increased potential for intense stand replacement fire, which was not historically characteristic

FIGURE 22.3. Characteristic old-growth ponderosa pine forest showing the dominance of large, old pine trees and overall low stand density. (Photo: J. Franklin)

of these sites, and (ii) continued reductions in residual populations of old-growth pines due to bark beetle attack, partially due to increased competition from younger trees. These outcomes are not consistent with current societal goals to sustain ecological and other social values in dry forest landscapes, which typically emphasize old-growth trees.

Consequently, programs are under way to restore more characteristic and sustainable conditions in many dry forest sites (Franklin and Agee, 2003). Although such silvicultural activities are logical, they are still controversial. Some scientists argue that high-severity fires were widespread in western pine forests and, therefore, should not be viewed as uncharacteristic. Some environmentalists argue that wildfire is preferable to human activities that modify stand structure and fuels.

Northern spotted owls further complicate management in some eastern Cascade Range mixed-conifer forests. The owl and its major prey, the northern flying squirrel, require denser forest patches that are potentially at risk of high-severity fire; silvicultural activities in such stands would alter existing owl and squirrel habitat. This conflict has impeded activities and resulted in losses of habitat to large, high-severity fires, such as occurred on the Sisters District of the Deschutes National Forest in 2003.

FIGURE 22.4. Eastside old-growth forest on a dry mixed-conifer site illustrating the uncharacteristic accumulation of fuel as a result of fire suppression programs. (Photo: J. Franklin)

Guidelines for Restoration of Old Growth on Dry Forest Sites

I believe that active management of old growth on dry forest sites is needed, despite the controversy. We do not want to continue to lose existing high-quality old-growth pine and dry mixed-conifer forests to high-severity fires, including what remains of suitable northern spotted owl habitat. Loss of irreplaceable residual old-growth ponderosa pine trees to competition-induced beetle mortality is also undesirable.

Active silvicultural management to reduce fuel loadings and reduce competition is imperative to conserve existing populations of old-growth trees and, ultimately, to restore forest structure and the populations of old-growth trees to sustainable conditions in an increasingly fire-prone climate. Prescribed fire is also an important silvicultural tool in restoring and maintaining appropriate fuel and seedbed conditions and maintaining fire-dependent ecological processes and biota.

Treating fuels and competing vegetation around old-growth trees is an appropriate starting point for silvicultural activities in dry forest stands (Johnson et al., 2007). Although old trees are most characteristically ponderosa pine, other species, such as western larch, sugar pine, or Douglas-fir, may be present. These old trees are structural keystones of dry forest ecosystems and practically irreplaceable; they provide critical habitat as living trees, snags, and logs. Old trees are most likely to survive wildfire and facilitate subsequent recovery. Yet, they are at risk of elimination by high-severity fire and competition-induced insect attack in most existing stands.

After addressing risks to old-growth trees, silvicultural activities can focus on stand-level goals, such as reducing fuels, reducing stand basal area, increasing mean stand diameter, shifting composition toward more fire- and drought-tolerant species, and managing mature tree populations to provide replacement old-growth trees. Targets for these parameters will vary with the plant association. Ground and ladder fuels should be the primary focus of fuel treatments (Franklin and Agee, 2003). Because pine and dry mixed-conifer old-growth stands are characteristically structural mosaics, silvicultural prescriptions should respond to and reinforce existing spatial heterogeneity, such as by retaining patches of seedlings, saplings, and small poles.

Restoration activities must be planned and implemented at the landscape scale for effectiveness. Landscapes characterized by low-severity fire have been dramatically transformed by human activities into landscapes of dense, continuous forest cover (Hessburg et al., 2005)—perfect conditions for watershed-level wildfires. Fuel treatments must be done at scales and in patterns that reduce the potential for extensive high-severity fire events; treating isolated stands is not likely to be effective.

Landscape-level fuel treatments are particularly critical where maintaining northern spotted owl habitat is an objective. Dense stands of more shade-tolerant species, such as grand fir, provide essential nesting, roosting, and foraging habitat for the owl and its prey; these are the stands most at risk of stand-replacement fire. Such habitats are best maintained by embedding them as islands in landscape matrices that have been treated to reduce the potential for high-severity wildfires.

Effects of Climate Change

Many years ago, several of us hypothesized that the Pacific Northwest has high vulnerability to climate change because of its summer-dry climate (Franklin et al., 1992). We also predicted that effects of climate change would be appear most profoundly as altered forest disturbance regimes, such as wildfire and insect outbreaks. Recent studies appear to validate these predictions (e.g., Breshears et al., 2005; Westerling et al., 2006).

So, will climate change affect our approaches to conservation of old-growth forest? Probably it simply reinforces the approaches that we are already taking—saving what we have and restoring it where we would like more.

Management responses to the increasing risks to old-growth forests resulting from climate change will differ between moist and dry forest sites. On moist sites the focus is on old-growth forests and, perhaps, intensified efforts to control wildfires; reducing fuel loadings in moist-site old-growth forests is *not* an appropriate response, however, because large fuel (biomass) accumulations are essential ecological features of such forests. Furthermore, established old-growth forests should be able to tolerate considerable climate change (Franklin et al., 1992).

On dry forest sites, climate change encourages us to protect old-growth trees by restoring stands and landscapes to more sustainable conditions. Activities that improve survival of the key structural elements should be of highest priority, given recent wildfire history; furthermore, silvicultural targets should reflect the probability that fire seasons will be ever longer and more severe.

LITERATURE CITED

Breshears, D. B., N. S. Cobb, P. M. Rich, K. P. Price, C. D. Allen, R. G. Balice, W. H. Romme, et al. 2005. Regional vegetation die-off in response to global-change-type drought. *Proceedings of the National Academy of Sciences* 102(42):15144–48.

Franklin, J. F., and J. K. Agee. 2003. Forging a science-based national forest fire policy. *Issues in Science and Technology* 20(1):59–66.

Franklin, J. F., and C. T. Dyrness. 1988. *Natural vegetation of Oregon and Washington.* Corvallis: Oregon State University Press.

Franklin, J. F., and M. A. Hemstrom. 1981. Aspects of succession in the coniferous forests of the Pacific Northwest. In *Forest succession,* edited by D. C. West, H. H. Shugart, and D. B. Botkin. New York: Springer-Verlag.

Franklin, J. F., and R. Van Pelt. 2004. Spatial aspects of structural complexity in old-growth forests. *Journal of Forestry* 102(3):22–27.

Franklin, J. F., F. J. Swanson, M. E. Harmon, D. A. Perry, T. A. Spies, V. H. Dale, A. McKee, et al. 1992. Effects of global climatic change on forests in northwestern North America. *Northwest Environmental Journal* 7:233–54.

Franklin, J. F., T. A. Spies, R. Van Pelt, A. B. Carey, D. A. Thornburgh, D. R. Berg, D. B. Lindenmayer, et al. 2002. Disturbances and structural development of natural forest ecosystems with silvicultural implications, using Douglas-fir forests as an example. *Forest Ecology and Management* 155(1):399–423.

Hessburg, P. F., J. K. Agee, and J. F. Franklin. 2005. Dry forests and wildland fires of the inland northwest USA: Contrasting the landscape ecology of the pre-settlement and modern eras. *Forest Ecology and Management* 211:117–39.

Johnson, K. N., J. F. Franklin, and D. L. Johnson. 2007. *A plan for the Klamath Tribe's management of the Klamath Reservation Forest.* http://www.klamath-tribes.org/forestplan.htm.

Pacific Northwest Research Station. 2002. *Restoring complexity: Second-growth forests and habitat diversity.* Science Update 1. Portland, OR: USDA Forest Service, Pacific Northwest Research Station.

Poage, N. J., and J. C. Tappeiner II. 2002. Long-term patterns of diameter and basal area growth of old-growth Douglas-fir trees in western Oregon. *Canadian Journal of Forest Research* 32(1):1232–43.

Sensenig, T. S. 2002. *Development, fire history and current and past growth of old-growth and young-growth forest stands in the Cascade, Siskiyou, and mid-coast mountains of southwestern Oregon.* Ph.D. dissertation, Oregon State University, Corvallis.

Tappeiner, J. C., D. W. Huffman, D. Marshall, T. A. Spies, and J. D. Bailey. 1997. Density, ages, and growth rates in old-growth and young-growth forests in coastal Oregon. *Canadian Journal of Forest Research* 27(5):638–48.

Westerling, A. L., H. G. Hidalgo, D. R. Cayan, and T. W. Swetnam. 2006. Warming and earlier spring increase western U.S. forest wildfire activity. *Science* 313(5789):940–43.

Chapter 23

Managing Young Stands to Develop Old-Forest Characteristics

JOHN TAPPEINER

Silviculture is the science and art of managing forest vegetation to meet landowners' objectives (Tappeiner et al., 2007). These objectives might include wildlife habitat, wood production, and reduction of potential for severe fire. The practice of silviculture in forest management, just like the place of old-growth forests, has evolved during the three decades since the environmental laws of the 1970s. When forest management objectives change, silvicultural practices must adapt.

From the early 1940s until the early 1980s, silviculture research and practice focused on reforestation following severe fires and timber harvest. Silviculturists conducted studies on the process and effectiveness of natural regeneration. We learned to produce high-quality conifer seedlings in nurseries for planting and to control competing vegetation to successfully regenerate forests. Thinning has been done for many years in western forests, and its effects on tree growth and stand development (Tappeiner et al., 2007) are well documented. However, we need better information on the response of understory trees and shrubs to thinning and prescribed burning and on the development of multiaged stands.

Silviculture objectives are defined by management objectives, which can be met by many different structures of both young and old stands. The Northwest Forest Plan objective of producing more older forest characteristics on federal lands depends on thinning young stands to produce stands of large trees, multiple layers of trees, large *snags* (standing dead trees), large logs on the forest floor, and a cover of shrubs in the understory. These characteristics are desirable in some forests because they provide habitat for species such as the spotted owl, but they make stands susceptible to severe damage from fires in fire-prone forests. Thus, an important issue is how to balance management objectives. From a silvicultural perspective, it is not difficult to produce either the dense old-forest characteristics needed by owls or the open, fire-resistant old-forest structures, but the two can't be done simultaneously in the same stand.

Silviculturists must know their forests and the suite of site-specific agents that affect stand development. They include

- *Root diseases* that kill overstory trees
- *Swiss needle cast,* a foliar disease of Douglas-fir that, in northwest Oregon, clearly alters the potential for infected stands to achieve old-forest characteristics
- *Wind, ice, and snow* that break and uproot trees
- *Bark beetles* that kill conifers in high-density stands, especially during periods of drought
- *Mistletoe* that redues tree vigor and predisposes them to mortality from insects
- *Severe fire.*

These factors *can* be beneficial, because they provide snags, logs on the forest floor, nesting cavities, and forest openings. Silviculture prescriptions evaluate the possible effects of these variables on stand development and provide strategies that respond to them. Communicating these strategies to the public is an important task.

Young Forests in the Pacific Northwest

Today's forests in Washington, Oregon, and northern California contain millions of acres of young stands, started principally for high yields of wood and other uses. They are often single species, usually Douglas-fir, and quite dense (fig. 23.1). Young-growth management, in many forest types, has often meant growing a stand for wood production for forty to sixty plus years, possibly with some thinning, then harvesting and starting a new stand. However, for significant proportions of public lands, today's policy calls for growing trees for longer periods to produce stands with many of

FIGURE 23.1. An unthinned, ninety-year-old Douglas-fir stand in western Oregon. Trees (about 300 per acre) have smaller stem diameter, and crowns are shorter than in a comparable thinned stand (see fig. 23.2). There are no trees in the understory, and shrubs are smaller. The potential for developing a complex, old-growth-like stand is minimal. (Photo: J. Tappeiner)

the characteristics of old growth: large trees and a diverse understory of trees and shrubs, as·well as large snags and logs on the forest floor. In some areas, plans call for thinning stands to produce these characteristics and then letting them grow naturally. On other lands, the goal is to manage stands on long rotations (more than 120 years) and maintain old-forest characteristics for several decades before the stand is cut.

In forests in northern California, southwestern Oregon, and east of the Cascades, fire is a threat, not only to young stands but also to old-growth stands. In these forests, treating stands to reduce the potential for severe fire is at least as important as developing characteristics of old forests.

There are considerable differences among young stands, even those established as plantations. Differences are caused by factors such as species composition, site productivity and growth rates, climate, past management, and varying effects of disturbances such as insects and pathogens. Each stand must be treated based on its current conditions and its potential for development, and no single treatment should be applied throughout the region.

Understory development differs significantly with understory species present. Western hemlock is very common in the Cascades and western Washington, where it readily becomes established in the understory after thinning, but its occurrence is sporadic in central and southern Oregon coastal forests. Redcedar is also common but less widespread than hemlock. Bigleaf maple, tanoak, and Pacific madrone are common in the coastal and southern forests, and they form a second layer after thinning.

Shrub species vary throughout the region and can have important effects on stand development. Vine maple, salal, huckleberry, salmonberry, and other common understory shrubs in Washington and Oregon coastal and Cascade forests often form dense continuous covers that inhibit the establishment of trees in the understory. Ceanothus and manzanita species regenerate following disturbance in northern California and southern Oregon, inhibit establishment of trees in the understory, and become fuels that can carry fire into the overstory trees. However, understory species also provide forage and cover for wildlife. Stands managed to reduce the potential for a severe fire will have few trees and flammable shrubs in the understory.

Rates of tree growth vary substantially across Pacific Northwest forests. On productive sites, the larger trees become more than 120 to 130 feet tall in fifty years and more than 170–190 feet tall in 100 years; on less-productive sites, tree height may range from 80–90 feet in fifty years and 110–130 feet at 100 years. Thus, age alone is not a good predictor of stand development.

In the northern part of the region major fires occurred at intervals of about 100 years or more before European settlement. In the southern Cascades and Siskiyous, fires were much more frequent, occurring at intervals of about five to thirty years or more (Sensenig, 2002). However, fire has been suppressed in these forests for many decades. Consequently, understories of trees and shrubs have become established, increasing the probability of severe crown fires.

Ownership and management goals vary across the region, leaving their legacies in today's stands. Some stands were planted at low densities or thinned at early ages; others have grown at high density for several decades, with little understory present. Older (thirty plus years) stands that have been thinned have a conifer and shrub understory (Bailey and Tappeiner, 1998). Species composition can also be affected during reforestation, competition control, and early thinning. For example, removing or retaining hardwoods and shrubs will cause further changes in the development and future structure of both mixed- and single-species stands.

In southern Oregon shelterwood methods of regeneration (leaving large trees for seed and shade) and heavy thinning have resulted in stands with a mixture of tree sizes and species composition. With this type of stand structure, treatments will focus on maintaining the large trees and managing the understory to reduce fire potential in some stands and favoring a diverse but flammable structure for wildlife in others. With trees already present in the understory, future development will be very different from that of young, dense stands with a single layer.

This discussion may make it seem very difficult to manage young stands for old-forest characteristics. However, with local knowledge of the forest, experience gained from conducting reforestation and thinning operations, and evaluation of results from thinning, forest managers can design treatments to produce old-forest characteristics. The condition of the stands themselves can indicate the proper treatments. The density of many young stands is so high that a very simple light initial thinning, followed by stand development for ten to twenty years, will result in immediate ecological benefits and provide many more options for future treatments.

Reasons for Thinning for Old-Growth Characteristics

Would it be best to let young stands develop naturally into old-growth stands, to "let mother nature do it"? It is becoming increasingly clear that many old-growth stands did not begin as dense, single-species stands like plantations. We reconstructed the history of old-growth stands and, to our surprise, we found that on many sites in western Oregon:

- Tree ages in old-growth stands varied by more than 100 years.
- Large old-growth trees grew quite rapidly in diameter when they were young.
- Tree size at 200+ years was positively correlated with their growth rate at age fifty years (Tappeiner et al., 1997; Poage and Tappeiner, 2002; Sensenig, 2002).

We concluded that low density—in at least parts of a stand—from prolonged establishment, irregular spacing, or disturbance from wind or low-severity fire combined to enable rapid growth and development of large trees and diverse stands. Today's young stands generally have neither the variable spacing and areas of low density nor the prolonged establishment period of old-growth stands. In fact, state and federal regulations usually require prompt regeneration of dense stands, and indeed this is good policy for many forest management objectives.

Then there is always the danger of severe fire in many of these dense young forests on dry, fire-prone sites. Although not every young stand or stand with large amounts of fuel needs treatment, strategic fuel reduction across large areas would reduce the potential for severe fire in many forests and landscapes. Old stands too have a high potential for severe fire because of decades of fire suppression and also could benefit from reduction of fuels and stand density.

Effects of Thinning on Trees and Stands

Thinning or density reduction affects both the growth of individual trees and stand development (Tappeiner et al., 2007). Light and available soil water increase after thinning, and the remaining trees and shrubs increase foliage density and photosynthesis. This provides carbohydrates for increased growth of the main sprouts. Seed production by trees and shrubs (Tappeiner et al., 2007) is stimulated by increased tree and shrub vigor. In western Oregon, a common difference between thinned and unthinned stands has been the large increase in the numbers of natural seedlings in thinned stands, a beginning for multilayered stands (Bailey and Tappeiner, 1998; Bailey et al., 1998). In fact, regeneration of western hemlock in the understory can be so dense on some sites that it can shade out herbs and

FIGURE 23.2. A thinned ninety-year-old Douglas-fir stand (about fifty trees per acre) that has begun the development of a layered stand by enabling the growth of bigleaf maple and establishment of conifers in the understory. Additional reduction of stand density by thinning, wind damage, etc., would increase the growth of trees and shrubs in the understory and aid the development of a complex stand. (Photo: J. Tappeiner)

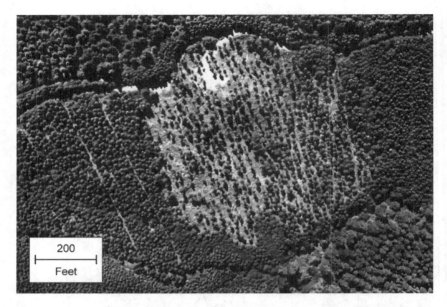

FIGURE 23.3. Aerial view of a recently thinned 25-30-year-old Douglas-fir stand with variable tree densities. Trees per acre after thinning range from 100–220 (far left) to 30 to 60 (center).

shrubs. Variable-density thinning may be needed in such cases to achieve a patchy understory (fig. 23.3). Shrubs such as salal can also dominate the understory. Managing for an understory beneath rapidly growing conifers is a relatively new practice; monitoring of both canopy density and understory growth will be needed to determine how best to promote multilay-ered old stands.

Hardwoods (bigleaf maple, California black oak, madrone, tanoak) can provide substrates for lichens and bryophytes and nesting cavities. Hardwood height growth rates are generally more rapid than those of conifers up to about twenty to thirty years of age, after which conifers grow more rapidly. Where the goal is to have vigorous, mast-producing (nuts and seeds eaten by animals) hardwoods for habitat, it will likely be necessary to reduce conifer density around hardwoods. Where a dense cover of shrubs interferes with the development of a conifer understory, reduction in shrub density and planting will likely be needed to establish understory trees. On fire-prone sites a tree and shrub understory can increase the ladder fuels and the potential for a severe fire, so developing such an understory will not be a widespread practice on these sites.

It is important to realize that thinning, planting, and other silvicultural practices alone will not achieve old-forest characteristics. Thinning will provide the space for trees to grow large crowns and stems. Once trees grow to large sizes, other factors such as stem breakage, decay, woodpeckers, and uprooting provide cavities that are nesting sites for birds, insects, and rodents, as well as substrate for fungi, lichens, bryophytes, and microorganisms that are all part of old forests. Silvicultural treatments provide trees and shrubs that can respond to natural disturbances and thereby provide old-forest characteristics.

Guidelines for Developing Old-Forest Characteristics

The major structural characteristics (large trees, dead trees standing and on the forest floor, shrub covers, and multiple tree layers) of old forest stands are relatively easy to achieve with a combination of treatments and natural processes. However, currently there are no agreed-upon future goals for young stands, with the exception of snags and wood on the forest floor. For example, what percentage of cover of what shrub species and what density, size, and arrangement of overstory and understory trees are desirable? Currently, forest managers are operating under edicts such as "more understory diversity is good" and "the greater the overstory density the better," but the time has come for clarification. Silviculturists and other forest managers could inventory local old-growth stands to determine their structure and use them as "models" for today's young stands to reach in the future. Some documented surveys of old-forest structure could serve to provide examples of data collection and analysis (Spies and Franklin, 1991; Poage and Tappeiner, 2005). An especially nasty problem is how to maintain these characteristics and reduce the potential for severe fire. Some of the structural characteristics of old-growth forest may be unnatural on fire-prone sites, because they developed with fire exclusion. Characteristics of old growth probably need to be redefined to include structure over large areas on fire-prone sites.

Current guidelines discourage managing stands that are more than eighty years old on federal land. The so-called eighty-year rule was established by the Northwest Forest Plan to focus attention on the need to manage younger stands. However, it has served its usefulness and inhibits managing stands to produce or protect old-forest stands in several ways:

- It does not allow fuel reduction that may be needed in fire-prone landscapes.

- Many acres of young stands will reach eighty years with no thinning because there are too many plantations for managers to treat in a timely way.
- A single thinning before age eighty may not be enough to enable young forest stands to develop desired old-forest structures.
- The inability of managers to rethin stands forces them to use a single heavy thinning that may unacceptably increase the risk of stand damage or produce an overly dense understory of shrubs and trees.
- Some stands are multiaged (size) with ages of larger trees ranging from less than fifty years to more than 200 years. What is stand age?
- Simply becoming eighty years of age does not ensure that a stand has or will develop desired levels of old-growth structure.
- Stand age does not take into account stand development, as noted above.

Planning a program of young stand management is an important part of the development of old-forest characteristics. Currently, a very small percentage of the young stands are being thinned, and it appears that a majority of them will reach eighty to 100 years or more with no thinning. Forest managers must weigh the options of (i) treating only very few stands with thorough, one-time treatments and severe reduction in density or (ii) doing a light thinning in many stands to begin the process of old-growth development and develop options for future treatments—but can they be implemented? The strict application of the eighty-year rule inhibits new approaches to forest management such as long rotations (Kohm and Franklin, 1997).

Need for Local Information

Recent research has expanded our understanding of treatments that will encourage old-growth forests and indicated relatively simple, site-specific information that will help provide goals for stand management. These include size, number, and species composition of overstory and understory trees, snags, and logs; shrub density; and gap size.

The following simple principles from research and observation generally apply:

- Keep large trees from the previous stand.
- Save less-common tree species depending upon stand species composition.
- Allow some disturbance to the forest floor by skidding logs or prescribed fire because it provides seed beds for tree and shrub seedling establishment.
- Control shrub and hardwood density as necessary to facilitate establishment and growth of natural and planted regeneration of conifers.
- Thin conifers from around hardwoods to help maintain hardwood vigor and mast production.
- Where prescribed burning occurs, avoid creation of a dense understory of shrubs that can become ladder fuels that carry fire to overstory conifers and also interfere with understory conifer establishment.
- Thin around selected large trees to maintain large crowns and continue rapid diameter growth and possibly mast production.
- Thin around patches of advanced regeneration of seedlings and saplings in the understory to develop multiple-story stands.
- Grow enough large trees to allow some to be killed to provide snags and logs.
- Maintain some dense patches of overstory trees to reduce shrub density and potentially promote establishment of small, herbaceous plants.
- Use stumps from old trees, where present, to help provide guidelines for tree numbers and arrangements when thinning young stands.
- Keep some parts of the stand unthinned or only lightly thinned as necessary to promote variation.
- Implement these guidelines with spatial variability and flexibility in mind: All objectives cannot and should not be achieved on each acre.

After thinning young stands, many old-growth characteristics will not occur for decades. However, early stand development can improve habitat for many species of birds as a shrubby understory develops (Hayes et al., 1997; Muir et al., 2002). Hardwoods that respond to thinning can produce substrate for lichens and mosses (Muir et al., 2002).

Can This Type of Thinning Be Economical?

Costs for thinning projects accrue from a variety of activities, including surveys to determine how key plants and animals are dispersed, protection of riparian areas, consultation with the public, marking trees to thin or leave using the guides suggested above, planning and contracting for timber removal, and preparing documents and records. A simple, effective record system is needed to ensure that managers and the public learn both short and long-term effects of the treatments. The value of timber thinned usually covers these costs. In today's log markets that favor small trees, sufficient wood can be removed, without compromising the objective of promoting old-forest characteristics, to pay the costs of thinning. An important aspect of a thinning program is to ensure a reasonably steady number of thinning projects so that experienced, capable, conscientious woods-workers earn living wages for their high-quality work.

Conclusion

Silviculture is crucial to managing today's young stands to become tomorrow's old forests and to creating fire-resistant old forests in fire-prone regions. The details of a program for managing young stands for old-forest characteristics will vary in response to differences in forest type, fire ecology, and other ecosystem variables throughout the Pacific Northwest. The silvicultural treatments needed are not complex, but an experienced workforce solidly trained in forest ecology and silviculture is essential to achieving these goals.

LITERATURE CITED

Bailey, J. D., C. Marysohn, P. Doescher, E. St. Pierre, and J. C. Tappeiner. 1998. Understory species in old and young forests in western Oregon. *Forest Ecology and Management* 112:289–302.

Bailey, J. D., and J. C. Tappeiner. 1998. Effects of thinning on structural development in 40- to 100-year-old Douglas-fir stands in western Oregon. *Forest Ecology and Management* 108:99–113.

Hayes, J. P., S. S. Chan, W. H. Emmingham, J. C. Tappeiner, L. D. Kellogg, and J. D. Bailey. 1997. Wildlife responses to thinning young forests in the Pacific Northwest. *Journal of Forestry* 95:28–33.

Kohm, K. A., and J. F. Franklin, editors. 1997. *Creating a forestry for the 21st century: The science of ecosystem management.* Washington, D.C.: Island Press.

Muir, P. S., R. L. Mattingly, J. C. Tappeiner II, J. D. Bailey, W. E. Elliot, J. C. Hagar, J. C. Miller, E. B. Peterson, and E. E. Starkey. 2002. *Managing for biodiversity in young Douglas-fir forests in western Oregon.* Biological Science Report USGS/BRD-2002-0006. Corvallis, OR: U.S. Geological Survey, Forest and Range Ecosystem Science Center.

Poage, N. J., and J. C. Tappeiner II. 2002. Long-term patterns of diameter and basal area growth of old-growth Douglas-fir trees in western Oregon. *Canadian Journal of Forest Research* 32(1):1232–43.

Poage, N. J., and J. C. Tappeiner. 2005. Tree species and size structure of old-growth Douglas-fir forests in central western Oregon. *Forest Ecology and Management* 204:329–43.

Sensenig, T. 2002. *Development, fire history, and current and past growth rates of old-growth and young-growth forest stands in the Cascade, Siskiyou, and mid-coast mountains of southwestern Oregon.* Ph.D. dissertation, Department of Forest Science, Oregon State University, Corvallis.

Spies, T. A., and J. F. Franklin. 1991. The structure of natural young, mature, and old-growth forests in Oregon and Washington. In *Wildlife and vegetation of unmanaged Douglas-fir forests*, technical contributors L. F. Ruggiero, K. B. Aubrey, A. B. Carey, and M. H. Huff. General Technical Report PNW-GTR-285. Portland, OR: USDA Forest Service, Pacific Northwest Research Station.

Tappeiner, J. C., D. W. Huffman, D. Marshall, T. A. Spies, and J. D. Bailey. 1997. Density, ages, and growth rates in old-growth and young-growth forests in coastal Oregon. *Canadian Journal of Forest Research* 27(5):638–48.

Tappeiner, II, J. C., D. A. Maguire, and T. B. Harrington. 2007. *Silviculture and ecology of western U.S. forests*. Corvallis: Oregon State University Press.

Chapter 24

Managing Forest Landscapes and Sustaining Old Growth

CHADWICK DEARING OLIVER

People value many things that forests provide—biodiversity, commodities, water quality, recreation, scenery, and protection from fires, among others. They want these values in forests throughout the world, and they want them at reasonable costs. Some of these values can be provided synergistically. Others are incompatible, so that supplying one value reduces the forest's ability to supply another. The challenge of management is to provide the maximum possible amounts of all values and, where values are incompatible, to offer acceptable trade-offs among them. Management also necessarily addresses less-than-ideal situations. For example, if sufficient areas of valued old-growth forest currently do not exist in an ecosystem, how can this situation be remedied?

This chapter examines how Pacific Northwest old growth can be maintained as one of the many values people want from the world's forests. The chapter first describes old-growth forests, explaining why they are valued, why there is confusion over their development, and why they are only one part of the natural forested landscape. It then discusses old growth in the Pacific Northwest as one value to be considered in a global context. Finally, it describes how Pacific Northwest old growth can be maintained using a triad approach.

Structures, Functioning, and Habitats
Provided by Old Growth

A dramatic new paradigm in ecology gained strength in the 1970s and is still percolating through the scientific world and the public: the realization that natural forests are open, dynamic systems. Accompanying the previous paradigm was the attitude "that nature knows best and that human intervention in it is bad by definition" (Stevens, 1990). Instead, forests are constantly changing through growth, advancing age, and disturbances.

There was concern that a rush to embrace the new paradigm might lead to assuming that all old forests could be harvested with no adverse consequences, because such harvests have many similarities to natural disturbances. But scientists embracing the new paradigm recognized that some stands of trees had not been completely destroyed for a long time; these stands provide unique values but do not behave as the presumed "climax" forests of the old paradigm. The term *old growth* was used to designate these old forests in a way that recognized their importance but avoided misinterpretations of the old paradigm.

In the late 1970s and early 1980s, two scientific papers inadvertently created confusion when they used "old growth" in different ways:

- Oliver (1981) described forest development in terms of dynamic, disturbance-oriented processes and referred to old growth as a very unusual, almost theoretical stage in the process that would occur if the forest grew for an extremely long time without being affected by a disturbance.
- Franklin et al. (1981) described old growth in terms of a stand's structure. The old-growth structure was described as having old, relatively large trees, several layers of tree canopies, dead trees and logs, and other distinguishing features.

The definitions are not compatible. Many stands in the old-growth structure (Franklin definition) achieved their condition through partial disturbances, a process different from that described by Oliver. Stands in the old-growth condition by Franklin's definition can be single and multiple cohort stands in the "stem exclusion" (stage when dense tree canopy suppresses tree regeneration), "understory," and "old-growth" stages described by Oliver. Consequently, "old-growth" can be extremely misleading if one assumes that the structure described by Franklin can be achieved only by the "process" of no intermediate disturbances described by Oliver.

At one time Franklin and Oliver discussed the confusion in the hallway of Anderson Hall at the University of Washington. There was agreement about the confusion but no immediate agreement about how to resolve it. Later, Oliver and Larson (1996) described the perplexity in the second edition of their book and tried to differentiate between Oliver's "process old growth" and Franklin's "structural old growth." Franklin et al. (2002) further elaborated on the distinction between the processes and structures and discussed the importance of woody snags and logs—structural features that have commonly been underemphasized. They also noted that much of this woody debris had been created before the stand was in the old-growth structure.

For identifying fauna and flora habitats, aesthetics, many ecological functions, and targets for active and/or passive management, Franklin's structural definition of old growth is more useful. Oliver's definition is primarily useful for understanding and managing the stand development process. For the remainder of this chapter, old growth will be discussed as a *structure* as described by Franklin (fig. 24.1). Old growth is sometimes referred to as the *complex structure,* although *complex* can also denote stands with most old-growth features but not necessarily old, large trees.

Stands can contain a variety of features that help classify them as old growth—large trees, woody snags, logs, multiple canopy layers, and others. These features do not necessarily occur together; instead, an old-growth stand can contain many areas with different combinations of these features.

Old Growth: One Structure in Dynamic Forest Landscapes

Old growth is one of several structures that a stand can be assigned to in various classification systems (see fig. 24.1; Franklin et al., 2002). Stands are constantly changing from one structure to another through growth, mortality, decay, and disturbances. A stand can change from the old-growth structure to the savanna or open structure by a fire, windstorm, insect or disease epidemic, or a combination of these and other events. By similar mechanisms, an understory structure can change to an old-growth structure. Even without a large disturbance, growth, advanced tree aging, death, disturbances, and regeneration may cause a stand to lose so many of its large, old trees or other features that its functionality as old growth is dimin-

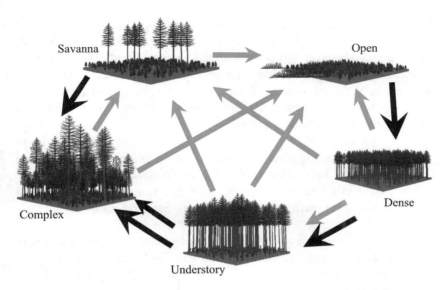

FIGURE 24.1. Each forest stand changes in structure along many possible pathways with growth, senescence, and disturbances. Black lines show changes created by growth, and gray lines show pathways created by disturbances. Each structure supports some species and other values. Other classifications exist in addition to those shown here. (© C. Oliver, Yale University)

ished, and it can have more characteristics of the dense, or another, structure. Growth and mortality can also cause a savanna structure to change to an old-growth structure.

Both tree growth rates and probabilities of disturbances vary with topographic position, so there is a higher probability that old growth is found in some areas than others (Camp et al., 1997). Before Euro-American settlement altered the landscape, catastrophic disturbances and regrowth likely caused fluctuations in proportions of different structures across the landscape (Oliver et al., 1998; Nonaka and Spies, 2005). The fluctuations were occurring at the same time that species were migrating in response to the changing climate, so an assemblage of old growth today consists of species that were not necessarily together in past millennia. Most species in the Pacific Northwest have migrated from farther south and are probably relatively resilient rather than highly sensitive to variations in habitat.

A major and continually contentious issue is whether and how to manage old growth. On the one hand, it could be argued that people are a part of nature, can create the same structural features that natural processes have created, and should be allowed to harvest and regrow old growth. The flaw here is the presumption that people have identified all important old-growth structures and also understand and mimic all processes that have created them. On the other hand, it could be argued that any level of management innately destroys the "naturalness," or purity, of old growth. This argument leaves the forest manager in a dilemma if there is not sufficient old-growth structure within a forest:

- Does the manager actively manipulate other forest stages to create as many old-growth structural features as possible?
- Or does the manager simply set aside forests that are not old growth and wait for them to grow, possibly, to old growth?

Active management would provide more of some old-growth features in the short term and thus relieve a near-term issue. Setting forests aside may provide old growth that has not been touched by people, but at a much later date.

Pacific Northwest Old Growth in a Global Context

Globally, people have made the moral decision that all species should be sustained—protected from extinction. A robust approach to implementing this decision is emerging as "sustainable forestry." Sustainable forestry has been described as an expansion of the Brundtland concept of "sustainable development" to include equity through both space and time: "People living in one place and time should provide their 'fair share' of values—neither unfairly exploiting nor depriving themselves of certain values to the detriment or benefit of people in another place or time" (Oliver, 2003).

Because of differences in the world's ecosystems, fair share does not mean equal or proportional sharing of values among ecosystems. Oliver (2003) and Oliver and Deal (2006) described a way to divide the world's forests into ecosystems and implement the "fair-share" principle using available data. The process identifies and addresses those ecosystems at the high and low extremes of vulnerability and exploitation and focuses appropriate actions.

Forests of the Pacific Northwest United States could be considered moderately endangered ecosystems. On the one hand, the total forest area is quite small; on the other hand, the forests are relatively well protected and contain relatively few species. For example, the Pacific Northwest "ecosystem" ("North and Central America temperate Oceanic"—ecoregions subdivided by continent [FAO 2001]) ranks in the smallest ten percent in terms of the world's forest area, with the largest forested ecosystem's area being nearly 300 times as large ("South American Tropical Rainforest," FAO, 2001). On the other hand, Oregon and Washington, the heart of the Pacific Northwest, contain 12.7 percent of their forests in protected areas (wilderness areas and parks; Haynes, 2003), compared to a world average of 12.4 percent. More than 50 percent of the forests in the Pacific Northwest would be protected if all national forests were included.

The Pacific Northwest has a relatively low number and concentration of species and endangered species compared to the southeastern United States and to California and an extremely low number and concentration compared to many tropical forests. The timber harvest rate is moderate, with the Pacific Northwest harvesting only 1.1 percent of its standing volume each year, compared to a U.S. average of 1.9 percent and a world average of 0.9 percent (Oliver and Deal, 2006). The range for all countries is less than 0.001 percent to more than 20 percent. Forests of the Pacific Northwest grow much more rapidly than most forests in the world.

Sustainable forestry management decisions are trade-offs among values within an ecosystem and among ecosystems throughout the world. Management's role is to provide as much of all values—as few negative trade-offs—as possible. For example, is it better to protect more old-growth forests in the Pacific Northwest and expect timber to be supplied from intensive plantations of exotic species grown on abandoned farmland or harvested forest land, for example, in Chile or Indonesia, or to reforest the tropical areas to native species and provide more timber from the Pacific Northwest? Analyses by Perez-Garcia (1993) have shown how people would shift timber extraction to other ecosystems (Perez-Garcia, 1993) and use more CO_2-generating steel, aluminum, concrete, and brick instead of wood if timber harvest were curtailed in the Pacific Northwest.

Fortunately, because much more wood is being grown in the world than is being consumed, a fair-share approach to the trade-off issue described above could still provide considerable areas for old growth in each ecosystem.

On the other hand, the forest areas in nearly all ecosystems have declined because of agriculture and other clearing. Thus, managers still must try to provide all forest values on smaller areas than were forested in the past. One challenge, for example, is to maintain all structures (fig. 24.1) across the landscape within each ecosystem, but an excess of one structure means too little of another structure, with a reduction of the values that it provides.

Each structure has unique management challenges, and a challenge of old growth is that it takes a long time to regrow if eliminated through natural or human disturbances. In such a case, old-growth-dependent species could become extinct or disappear from an area before old growth regrows.

Managing Old Growth as Part of a "Triad"

A robust way to provide all structures and their values is the triad system (fig. 24.2; Seymour and Hunter, 1999). The Pacific Northwest and many other forest areas of the world have been gradually developing toward this triad. The triad assumes each forested ecosystem can be divided into three zones, and the amount, placement, and treatment of each of these zones can be given further consideration:

1) *Intensive plantations, where forests are managed very intensively for commodities, usually timber.* Many practitioners and researchers have suggested that the world's wood be obtained from intensive plantations and the remaining forest be put into protected areas. Only ten percent of the world forest area would be required to meet the present and near-future timber needs; however, such an arrangement would be socially harmful to rural people, ecologically problematic, and financially very difficult for the timber industry (Oliver, 1999; Oliver and Mesznik, 2006). It is improbable that intensive plantations will be the sole source of timber and other forest products in the future.

2) *Protected areas, where forests are not managed, but "natural" processes without human interference are allowed to proceed.* Conceptually, protected areas would contain a variety of structures as disturbances, regrowth, and advanced aging cause change. Such protected areas may be the best place for old-growth forests if people want such forests to develop and exist without human intervention. In practice, judicious placement of protected areas could ensure that a preponderance of old-growth structure existed in them (fig. 24.2).

Several management issues would need to be addressed relative to old growth and these protected areas:

• Protecting or providing old growth needs to be examined in light of what types of forests currently exist—and *can* exist. For example,

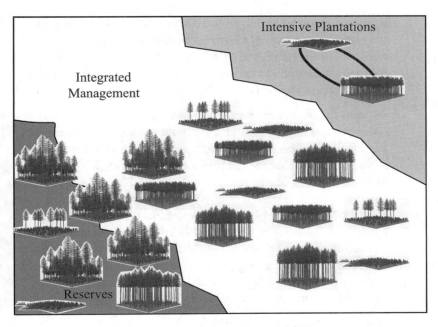

FIGURE 24.2. The triad approach for providing a spatial and temporal fair share of values in each ecosystem (Seymour and Hunter, 1999). Part of each ecosystem would be treated as a protected area, or reserve; part would be actively managed for multiple values, including all habitats; and part could be used for intensive plantations, if desired. (Figure from Oliver, 2003; © C. Oliver, Yale University)

if there are insufficient stands presently in the old-growth structure, then protecting or actively managing young forests to provide at least some old-growth structures as soon as possible could be prudent. However, it would not be scientifically defensible to remove the old forests and claim the young forests can be manipulated to provide all of the features (Subcommittee on Forests, Family Farms, and Energy, 1992). The decision about which stands should receive this manipulation can be assessed using "risk analysis" (Oliver, 1998).

- Because the forest is dynamic, forests in the protected areas may be destroyed through natural disturbances (e.g., volcanic eruptions, windstorms, or fires such as those recently affecting the Pacific Northwest) or may simply lose their old-growth features and functionality through time. It may be appropriate for stands in the "integrated management" area (see below) that are old and contain many old-growth features to be redesignated as protected areas. At the same time, stands not having the old-growth structure may be removed from the protected area and the timber salvaged to pay for the new protected area and to ensure that the Pacific Northwest ecosystem provides its fair share of timber.

3) *Integrated management areas, where forests are managed for a variety of commodity and noncommodity values.* In these areas, silvicultural activities can be used to mimic, avoid, and recover from natural disturbances to ensure that the full suite of stand structures is maintained. Maintenance of these structures provides timber and employment, and the diversity of structures provides fire protection, aesthetics, and habitats. Maintaining old-growth structures is part of "coarse-filter" biodiversity protection (Hunter, 1990)—ensuring that species do not become extinct by providing all habitats within an ecosystem.

These integrated management areas may be managed in concert with protected areas. Some stands in the integrated areas may be managed on long rotations for several reasons:

- To supplement old growth in protected areas;
- To enhance spatial continuity of appropriate, closed forest habitats between old-growth stands in protected areas;
- To provide potential replacement in case old-growth stands in protected areas are destroyed or otherwise lose their functionality; and
- To provide high-quality timber if or when these stands are harvested (as other stands grow to an old age).

Certain features could be planned in these integrated areas that make them more suitable to wildlife habitats and to potentially supplementing, or converting to, old growth in protected areas. Such features include a mixture of species, a diversity of tree spacing, and standing and downed woody debris, as well as older, living trees. To ensure the full diversity of the forest, such features could be provided in some places but not others.

Various incentive mechanisms have been suggested for encouraging private landowners to provide understory and old-growth structures in the integrated management areas (Lippke and Oliver, 1993; Lippke and Fretwell, 1997). These mechanisms are part of a growing movement to compensate forest landowners for providing "ecosystem services" (Heal, 2000; Wunder, 2005; Sandborn and Tyrrell, 2006).

The integrated approach has been suggested by members of the timber industry in the Pacific Northwest, who offered to manage their forests to provide the "early- and mid-successional" (e.g., open and dense; fig. 24.1) structures, expecting federal lands to provide the old-growth structures (Anonymous, 1994).

The questions raised by consideration of old-growth forests require value decisions and entail trade-offs: among biodiversity and other values, among different parts of the world, and among levels of risk in an uncertain world. Unlike the "climax" perspective, the more dynamic view of nature requires a greater understanding and appreciation of the dynamic nature of forests.

LITERATURE CITED

Anonymous. 1994. Ecological objectives for industrial forestlands: A discussion paper. *Journal of Forestry* 92(5):18–19.

Camp, A., C. Oliver, P. Hessburg, and R. Everett. 1997. Predicting late-successional fire refugia pre-dating European settlement in the Wenatchee Mountains. *Forest Ecology and Management* 95:63–77.

FAO (United Nations Food and Agriculture Organization). 2001. *Global forest resources assessment 2000*. Main report. FAO Forestry Paper 140. Rome: Food and Agriculture Organization of the United Nations.

Franklin, J. F., K. Cromack, Jr., W. Denison, A. McKee, C. Maser, J. Sedell, F. Swanson, and G. Juday. 1981. *Ecological characteristics of old-growth Douglas-fir forests*. General Technical Report PNW-GTR-118. Portland, OR: USDA Forest Service, Pacific Northwest Research Station.

Franklin, J. F., T. A. Spies, R. Van Pelt, A. B. Carey, D. A. Thornburgh, D. R. Berg, D. B. Lindenmayer, et al. 2002. Disturbances and structural development of natural forest ecosystems with silvicultural implications, using Douglas-fir forests as an example. *Forest Ecology and Management* 155(1):399–423.

Haynes, R. W., tech. coord. 2003. *An analysis of the timber situation in the United States: 1952 to 2050—A technical document supporting the 2000 USDA Forest Service RPA assessment*. General Technical Report PNW-GTR-560. Portland, OR: USDA Forest Service, Pacific Northwest Research Station.

Heal, G. 2000. *Nature and the marketplace: Capturing the value of ecosystem services*. Washington, D.C.: Island Press.

Hunter, M. L., Jr. 1990. *Wildlife, forests, and forestry*. Englewood Cliffs, NJ: Regents/Prentice Hall.

Lippke, B., and H. L. Fretwell. 1997. The market incentive for biodiversity. *Journal of Forestry* 95:4–7.

Lippke, B., and C. D. Oliver. 1993. A proposal for the Pacific Northwest: Managing for multiple values. *Journal of Forestry* 91:14–18.

Nonaka, E., and T. A. Spies. 2005. Historical range of variability in landscape structure: A simulation study in Oregon, USA. *Ecological Applications* 15(5): 1727–46.

Oliver, C. D. 1981. Forest development in North America following major disturbances. *Forest Ecology and Management* 3(1980–81):153–68.

Oliver, C. D. 1998. Passive versus active forest management. In *Forest policy: Ready for renaissance*, edited by J. M. Calhoun. Proceedings of a conference held at the Olympic Natural Resources Center. Contribution No. 78. Seattle: University of Washington, College of Forest Resources, Institute of Forest Resources.

Oliver, C. D. 1999. The future of the forest management industry: Highly mechanized plantations and reserves or a knowledge-intensive integrated approach? *Forestry Chronicle* 75(2):229–45.

Oliver, C. D. 2003. Sustainable forestry: What is it? How do we achieve it? *Journal of Forestry* 101(5):8–14.

Oliver, C. D., and B. C. Larson. 1996. *Forest stand dynamics,* update ed. New York: John Wiley & Sons.

Oliver, C. D., A. Camp, and A. Osawa. 1998. Forest dynamics and resulting animal and plant population changes at the stand and landscape levels. *Journal of Sustainable Forestry* 6(3/4):281–312.

Oliver, C. D., and R. L. Deal. 2006. A working definition of sustainable forestry and means of achieving it at different spatial scales. *Journal of Sustainable Forestry* 24(2/3):141–63.

Oliver, C. D., and R. Mesznik. 2006. Investing in forestry: Opportunities and pitfalls of intensive plantations and other alternatives. *Journal of Sustainable Forestry* 21(4):97–111.

Perez-Garcia, J. M. 1993. *Global forestry impacts of reducing softwood supplies from North America.* CINTRAFOR Working Paper 43. Seattle: University of Washington, College of Forest Resources, Center for International Trade in Forest Products.

Sandborn, R., and M. Tyrrell. 2006. Markets and payments for ecosystem services. *Yale Forest Forum Review* 9(1). New Haven, CT: Global Institute of Sustainable Forestry, School of Forestry and Environmental Studies, Yale University.

Seymour, R. S., and M. L. Hunter, Jr. 1999. Principles of ecological forestry. In *Maintaining biodiversity in forest ecosystems*, edited by M. L. Hunter, Jr. Cambridge: Cambridge University Press.

Stevens, W. K. 1990. New eye on nature: The real constant is eternal turmoil. *New York Times*. July 31, 1990. B-5–B-6.

Subcommittee on Forests, Family Farms, and Energy. 1992. *Review of high-quality forestry and extended timber harvest rotations*. Hearing before the Subcommittee on Forests, Family Farms, and Energy of the Committee on Agriculture, House of Representatives, 102nd Cong., 2d sess., March 11, 1992. Washington, D.C.: U.S. Government Printing Office.

Wunder, S. 2005. *Payments for environmental services: Some nuts and bolts*. CIFOR Occasional Paper No. 42. Indonesia: Center for International Forestry Research.

Chapter 25

Unexplored Potential of Northwest Forests

BETTINA VON HAGEN

The Northwest Forest Plan did not solve the dilemma of how to manage our extraordinary forest resource; it simply shifted the debate. Of course, the continuing distrust and friction between much of the environmental and forest industry communities and between rural and urban residents have serious implications for the region's citizens' ability to forge coherent, productive, long-term strategies for forest management. Distrust and friction also relate directly to broader societal concerns such as education, land use, and economic development. Conflicts over harvesting after fire on federal lands, the appropriate management of state lands, regulation of private forest lands, and other forest management issues continue to polarize us and have left bitter, unresolved legacies.

Underlying arguments about forests, of course, are deeper divides. The appropriate management of forests is a proxy for bigger rifts and social equity issues, such as the loss of political and economic power of rural communities to an increasingly urbanized population, or the ethic of those who work and live off the land versus those for whom it serves as a recreational amenity. Issues of class, economic opportunity, relationship to the land, legacy, and equity are at the heart of our debates. Ultimately, our forest management solutions, if they are to endure and build true value, must address these underlying tensions.

The irony in this conflict is that in the productive forests of the Pacific Northwest, it is not necessary to choose between fiber production and public values. To the contrary, this region is particularly well suited for a forest management approach in which sawlogs, pulp, jobs, carbon storage, flood control, habitat, scenic vistas, and recreational opportunities can all be produced from the same acres. Although many nonindustrial, state and municipal, tribal, conservation, community, and industrial forest landowners practice this approach, this ecosystem-based forestry approach constitutes the missing voice in the forest management debate.

Effects of the Northwest Forest Plan

Few of the world's temperate forested regions have witnessed such protracted and acrimonious debate about their future and management as the Pacific Northwest. The Northwest Forest Plan (see box 2.1) successfully addressed the immediate issue of logging old-growth forests on public land, but it also generated sharply bifurcated management strategies on federal forests (emphasis on biodiversity) and private forests (emphasis on fiber maximization). It also failed to address some key realities.

First, the plan created late-successional reserves that could not provide all the public goods required. Second—and most importantly—it did not address the role that private lands could play in contributing to public values such as recreation, scenic vistas, carbon storage, habitat, and clean water, all while producing timber. By creating no incentives for forest management on private lands to enhance public values and by allocating insufficient resources for public lands to provide these, the plan took a limited view of the goods and services we desire from our forest landscape and thus spawned continuing controversy and debate.

Fortunately, there is a new market emerging that provides another opportunity for private forest landowners to implement a different approach to forest management, focusing not only on timber production but also on the enhancement of public goods. Ecosystem service markets provide an opportunity to achieve environmental goals through private instruments, such as the use of carbon credits to reduce greenhouse gas emissions (box 25.1). The advent of new incentives for conservation is particularly timely, because private forest land is rapidly changing hands in the Pacific Northwest and elsewhere, with about half of the land previously held by integrated forest product companies now owned by financial funds. Many fear this may lead to greater management intensity and forest fragmentation.

BOX 25.1. A SAMPLING OF ECOSYSTEM SERVICE AND CONSERVATION FINANCE MARKETS RELEVANT TO FORESTS

Bettina von Hagen

- *Carbon credit* trading provides a market mechanism to meet greenhouse gas reduction targets by trading permits to emit carbon dioxide and other greenhouse gases. It is one way that countries, industry, and other regulated entities can meet their reduction obligations under the Kyoto Protocol and other regulatory and voluntary schemes. In addition to trading among regulated entities, emitters can invest in carbon sequestration strategies such as reforestation or forest management that sequesters additional carbon. While the United States is not a signatory to Kyoto, regional markets are emerging, such as the Regional Greenhouse Gas Initiative in the Northeast and the Western Climate Initiative, that are setting regional caps and establishing trading systems.

- *Conservation banks* are permanently protected privately or publicly owned lands that are managed for endangered, threatened, and other at-risk species. In exchange for permanently protecting the bank lands and managing them for listed and other at-risk species, conservation bank owners may sell credits to developers or others who need to compensate for the environmental impacts of their projects. Conservation banks are widely used in California and are in early development in other states. Forest applications to date include development of conservation banks for red-cockaded woodpeckers.

- A *wetland bank* is a mechanism for compensating for unavoidable impacts to aquatic resources permitted under Section 404 of the Clean Water Act by creating, restoring, or enhancing wetlands, streams, or other aquatic resources. Wetland banks can lead to larger and more ecologically beneficial wetlands than on-site mitigation while also reducing planning and management costs.

- *Water quality trading* is a market approach to meeting water quality goals mandated by the Clean Water Act. Under this approach, one source can meet its regulatory obligations by using pollutant reductions created by another source that has lower pollution control costs. Traded "pollutants" include nitrogen, phosphorus, sediment, and temperature. Tree planting has played a large role in systems requiring temperature reduction.

- *Conservation easements* are voluntary, legally binding agreements between a landowner and a land trust or government agency that permanently limit uses of the land to protect its conservation values. In the most typical case, a forest landowner permanently forsakes development rights, although working forest easements that restrict or define specific forest management activities or outcomes are becoming more common. Easements may be donated or sold, generally to public agencies. In 2006, the federal tax deduction provisions were significantly expanded.

- *Transferable development rights* allow landowners to purchase and sell residential development rights from lands that provide a public benefit. Such lands include forests, as well as farms, open space, and important habitat. Landowners receive financial compensation to forego development, and the public receives permanent preservation of the land. Transferred development, rights can be used to build additional houses on other parcels in more appropriate areas.
- *New Market Tax Credits* were created and authorized by the U.S. Department of the Treasury to stimulate investment in communities suffering from poverty and unemployment. Community development entities compete for a New Market Tax Credit allocation annually and are awarded an allocation based on their experience, capacity, and programs to stimulate investment and jobs in needy communities. An allocation gives the community development entity the capacity to sell a tax credit equal to thirty-nine percent of the qualified investment to a buyer, usually a financial institution, that can apply the credit to their federal tax burden.

Although the Northwest Forest Plan was successful in many ways, it also had some less-positive impacts, including lower timber harvests, fewer jobs than anticipated, perverse incentives, and unforeseen consequences. First, most of the burden for endangered species recovery fell on federal lands, setting up sharply different forest management approaches on federal and private lands. With the decline of timber harvests and revenue generation on federal forests, annual budgets fell as well. Funding declined for much-needed thinning of overstocked stands on federal forests. Some environmental groups continued to suspect that federal forest thinning sales were veiled attempts to cut more big trees and so challenged timber sales under the new program. Coupled with reduced management staff and resources, timber production from federal forests has been considerably lower than even the plan's sharply reduced targets foresaw.

Timber production shifted to the younger forests on private lands, and many of the mills tooled for the larger logs previously produced by federal forests closed. These events, coupled with the rise of engineered wood and the collapse of Asian markets, which had paid a premium for large logs, accelerated the transition to lower rotation ages and a smaller number of very large, highly automated mills, mostly located on transportation corridors away from traditional forest communities. As transportation distances to the few remaining mills capable of sawing larger logs have increased, the former premium paid for large logs has become a penalty. Many private forest landowners, some of whom might otherwise consider managing for longer rotations, have chosen to shorten rotations to reduce the risk of lower prices.

Daunted by the scope and breadth of the plan's impact on federal forest management and an increasingly urbanized population in the region with a hunger for forest amenities, private forest landowners also anticipate increasing regulation of private forest land. In response, many of them are wary of managing forests to attract endangered species, create recreational opportunities, or develop scenic vistas, for fear that these public interests will trump their ability to manage in accordance with their own private interests. This fear—perhaps exaggerated by some—has, along with many other regional and global factors, accelerated the transition to more-intensive, fiber-maximizing management on private lands.

We now have two highly differentiated forest management strategies: (i) federal forests managed for public values, albeit with few resources, and (ii) private forests managed intensively for fiber production. The latter seek to intensify further to compete with the fast-growing plantations and highly automated lumber processing of Chile, Brazil, New Zealand, and soon, China.

Management of state-owned forests in Oregon and Washington generally follows a middle course, producing timber and nontimber products and services, but is subject to continuing pressure from the timber industry and counties dependent on timber revenues to accelerate the cut—and from environmentalists to reduce it. In pursuing the sharply divided strategies of intensive production on private lands and biodiversity on federal lands, we have reinforced old, unproductive stereotypes: conservation versus development, environment versus jobs, timber production versus biodiversity protection, consumption versus conservation, and urban versus rural—false choices that undermine the creation of long-term societal value.

An Alternative Approach: Coproduction of Timber and Ecosystem Services

At Ecotrust—a regional conservation organization based in Portland, Oregon—we look for "triple-bottom-line" approaches, meaning that development and conservation strategies must (i) be financially viable, (ii) contribute to healthy and intact landscapes, and (iii) help build vibrant communities. We believe that all economic activity can and should meet this triple-bottom-line test, from managing forests and farms, constructing buildings, and transportation systems to manufacturing widgets. We do believe in the value of preservation—of protecting viable, representative areas of significant ecosystems, such as old-growth forests. In some cases, protection is the best triple-bottom-line option.

However, in the face of growing population and increased resource demands, protection strategies are insufficient to maintain ecosystem health over the long term and are often prohibitively expensive. How we manage the land from which we derive our food, fuel, and fiber—the commodity production lands—is ultimately much more critical to our long-term prosperity.

Moreover, there are few better places to test this approach than in the Pacific Northwest's productive, lush, and forgiving forest landscape. Unlike many timber-producing countries, such as South Africa, New Zealand, or Brazil, that rely on exotic plantations, our native species are highly desirable commercial species. Unlike most tropical forests, our native tree species diversity is fairly low, simplifying management and commercialization. Our land tenure is secure; our forest products industry—logging, processing, distributing—is efficient; our population is relatively small and prosperous. Our trees tend to grow tall, straight, and old, often not reaching the culmination of mean annual increment until seventy or eighty years or even beyond that with appropriate thinning (Curtis, 1997). Given this, and the persistence of snags, downed logs, and belowground biomass, our forests store more carbon than just about any other terrestrial ecosystem (Smithwick et al., 2002). Equally important, our forests have salmon. Commercially valuable, elusive, iconic, beautiful, and culturally significant, salmon capture nutrients from the ocean and deliver them to our forests' doorstep, enriching forest health and sustaining hundreds of plant and animal species (Cederholm et al., 2000).

Why not manage for all of these values, not just on a fraction of the landscape, but as a dominant forest management strategy? Why not explicitly manage for logs; for pulp; for biomass; for carbon; for habitat; for fish; for clean, cold water; for recreational opportunities; for scenic vistas on all of our private lands? Why not manage for older forests with the structure, diversity, and productivity to deliver not only timber but a broad array of nontimber products and ecosystem services?

At this point, one might reasonably ask, If such an approach is financially feasible, why hasn't it become the dominant management paradigm? Doesn't managing for values other than timber sharply decrease timber harvests, profits, and jobs? This seems reasonable but is only partly true and is mostly wrong. Many modelers and researchers have tackled the question of the relative financial, ecological, and social performance of different management strategies, with rotation length being a key variable and often serving as a proxy for habitat quality and structure. The findings suggest that managing forests for structure and diversity, or ecological forestry, results in (i) almost as much wood (or, in some simulations, as much wood) as industrial forestry; (ii) more valuable timber, due to both longer rotations and thinning for log quality; (iii) more jobs, given that thinning is labor-intensive and requires more frequent entries than industrial forest practices; and (iv) improved ecological outcomes. There is, however, considerable variability in the projected financial performance of long-rotation forestry, with results varying greatly depending on the rotation ages that are selected, the modeling assumptions, and the discount rate that is used.

One intriguing comparative study was conducted by Dr. Andrew Carey and his colleagues, who modeled three divergent forest landscape management strategies over a 300-year period in a Pacific Northwest coastal hemlock forest (Carey et al., 1999). The study concluded that the biodiversity pathway approach produced eighty-two percent of the net present value (the sum of the discounted net cash flows at the present time) of the industrial approach (a net present value of fifty-eight million dollars vs. the industrial approach's seventy million dollars) while achieving ninety-eight percent of the potential ecosystem health of unmanaged forests and also produced a larger variety and higher quality of wood products than the industrial approach. In a different study, Haynes (2005) found much larger differences in net present value. Not surprisingly, the "no-touch" approach generated no revenue; surprisingly, it failed to deliver even close to the level of ecological benefits provided by the biodiversity pathway approach. This illustrates that previously clearcut forests often benefit from active management to more quickly develop older forest characteristics.

Ecotrust's modeling generally confirmed Carey's (Carey et al., 1999) conclusions, although we found a larger difference in net present value between the industrial and ecological approach—thirty percent versus the eighteen percent difference noted by Carey, which we attributed to changes in prices for logs and pulp since Carey's study. In other words, a forest managed under the ecological regime produces more wood, more jobs, and more cash over time, but the harvests come later as rotation age is extended. Given the time value of money, distant cash flows are worth less than those closer in time, so the net present value is thirty percent lower (on average) for ecological forestry than for industrial forestry.

This was a tremendously exciting affirmation: If timber managed under ecological forestry could produce seventy percent of the industrial value, then the other forest products and services, such as carbon, biodiversity, and scenic values, which increase significantly under ecological management, could produce the other thirty percent of net present value to make up the difference. *This would allow ecological forestry to be fully competitive with industrial forestry from a financial investment perspective.* Given the recent rapid escalation of ecosystem service markets, such as carbon credits, conservation and wetland banking, and water quality trading, as well as longer-standing markets for conservation easements, we see expanding opportunities to monetize and transact in these other forest products and services (box 25.1). With a regional cap and trade market for carbon a virtual certainty in the next five years, along with the rapid growth of interest in socially responsible investing—now a $2.3 trillion market—we see an opportunity to develop more compelling forest land management and investment options.

There were other timely reasons to design and offer an alternative forest management fund: (i) poor returns in the stock market in the first few years of the twenty-first century were causing investors to seek alternative asset classes; (ii) increased recognition of the value of forest land investments as a hedge against inflation and as a portfolio diversification strategy; and (iii) as discussed below, a considerable increase in the amount of timberland changing hands, creating an opportunity for new forms of ownership and new entrants.

The region's integrated forest product companies, such as Boise Cascade, Crown Pacific, and Longview Fiber, have sold or are divesting of their forest lands either voluntarily or through unsolicited hostile takeover bids. The result is a rapid change of ownership from integrated companies to ownership of forest land by financial funds. Driven by tax considerations, excessive debt, the lackluster performance of forestry companies in the stock market, and an increase in financial capital seeking alternative investments, this transition has created a new class of forest land ownership. These new owners value forest land strictly for its financial characteristics—strong historical financial returns at relatively low risk, a hedge against inflation, and diversification from other portfolio assets—and often have weaker ties to forest communities than the integrated forest product companies they replaced.

Although the growing presence of *timber investment management organizations (TIMOs)* presents new risks, it also presents some new opportunities. Forest land is coming to market at an unprecedented rate, offering buying opportunities not only for traditional TIMOs but also for new classes of owners with genuine conservation, community, and tribal interests.

Ecotrust Forests LLC: Seeking Triple-Bottom-Line Returns

In 2004, Ecotrust created Ecotrust Forests LLC (the fund) to give investors an opportunity to own forests managed for the triple bottom line—competitive financial returns, improved forest health, and job generation in rural communities—using the TIMO structure as a template. Although we readily adopted the organizational structure of a TIMO, we had more difficulty assuming some of its other structural elements. TIMOs typically create funds with a ten- to fifteen-year life; capital is raised and placed into forest investments, and then the portfolio is liquidated (meaning the forest land is sold) at the end of the designated time period. This structure is fundamentally inconsistent with our management objectives, which are to

- Purchase industrial forest land and manage it for greater structural complexity, diversity, and long-term productivity
- Provide competitive returns for our investors through the production of high-quality timber and pulp and the monetization of ecosystem services such as carbon storage, habitat, and water quality (box 25.1)
- Concentrate land acquisitions in high-priority watersheds where our management can benefit salmonids and other species of interest
- Attempt to influence the entire watershed by colocating with other landowners that share our management objectives
- Create long-term relationships with local communities and contractors, providing a reliable stream of jobs and opportunities
- Expand the knowledge, understanding, and practice of managing commercial forests for the triple bottom line.

Our objectives demand a long-term—in essence, perpetual—ownership. Buying a young forest and building up the volume of high-quality trees and habitat takes time and serves no purpose if the property is to be sold after fifteen years, other than creating a forest with a high liquidation value. We considered reselling the property subject to a conservation easement that perpetuated our forest management but rejected this approach due to the limited funding of conservation easements in the western United

States, the paucity of conservation buyers, and the impossibility of creating the long-term relationships fundamental to community partnerships and research and monitoring efforts.

Outcomes of Long-Term Management

The result is a rather unusual and ambitious fund structure: Ecotrust Forests LLC continuously raises capital, purchases forest land, and manages those forests in perpetuity. Financial returns are generated through timber harvests and sales of nontimber forest products as well as ecosystem services. Exit opportunities are initially limited to private sales of membership interests to other qualified investors and to a limited buyback program of membership interests that starts in the tenth year. As the fund grows, we may consider changing the organizational structure from a limited liability company to a private or public real estate investment trust to improve liquidity opportunities for investors. Although timber revenues in the first two decades are generally lower than they would be under industrial management, nontimber revenues, including carbon credits, conservation easements, and New Market Tax Credits, are pursued in the first five years of ownership, providing a significant and early cash return to investors that can, in some cases, more than compensate for the delayed receipt of timber revenues.

As of October 2007, the fund managed 12,000 acres in four properties in coastal Oregon and Washington and had twenty-five investors, five of whom increased their capital contribution in the previous year. The fund is small but well poised for growth and enjoys considerable interest from potential investors. Investors can now see firsthand how the forest looks and responds to variable-density thinning (fig. 25.1) and how an aggressive focus on removing fish passage barriers improves access for salmon and overall habitat quality. We are also exploring and demonstrating the potential of alternative revenue streams; have successfully enhanced timber returns with conservation easements, sale of New Market Tax Credits, and special forest products (primarily leases for harvest of salal, used in the floral green trade, and salvaging old cedar stumps for production of shakes and shingles); and are structuring forest carbon projects for the voluntary carbon market (fig. 25.2).

Although these are early days, we have gained valuable insights from pursuing about a dozen acquisitions and successfully acquiring four properties, developing and implementing management plans, and seeking ecosystem service transactions. Some of the early lessons include

FIGURE 25.1. Dickey River postharvest, 2005. Aerial view of variable-density thinning and patch cuts. This approach to forest management enhances carbon stocks and biodiversity, which can be monetized to produce new revenue opportunities for forest landowners.

- The fund is most competitive in buying younger properties, because mature properties are priced at liquidation value, which favors buyers with aggressive logging plans. Younger properties also provide the opportunity to improve structure and diversity early in the stand's life by thinning.
- Ecological forestry is more sensitive to log and pulp prices than industrial forestry is. Thinning is more expensive than clearcutting on a volume basis (dollars per million board feet); the result is smaller net margins and potentially less profit if prices drop or if steep slopes or other factors (for example, distance from roads) increase costs. On the other hand, because the fund is not leveraged and fixed expenses are low, thinning and management expenses can be timed to match well-priced markets. In addition, because the fund has access to nontimber markets for ecosystem services, it can pursue these other revenue streams when log and pulp prices are low.

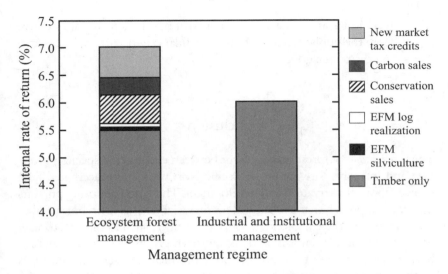

FIGURE 25.2. Although ecosystem-based forest management yields a lower internal rate of return when only timber harvests are considered, the sale of ecosystem services can enhance returns above those that are obtained by industrial forestry. (Binkley et al., 2006)

- Most growth and yield models and forest management systems are designed for the predominant industrial approach of clearcutting, site preparation, and planting. Finding modeling and management tools that adequately project growth and natural regeneration following thinning and small patch cuts is very challenging and reflects a significant underinvestment by the research community in forest management approaches that differ from the industrial model.
- Early evidence suggests that our commitment to long-term ownership, an explicit focus on producing jobs and opportunities for local communities, and a broader set of management activities produces more jobs and more-reliable employment than current industrial practices.
- Ecosystem service markets are a viable strategy for enhancing returns from timber and making ecological forestry fully competitive with the industrial model. The interest and opportunities in both carbon credit and water quality trading markets have increased significantly since the fund was formed (von Hagen and Burnett, 2006). Particularly significant has been the sale of New Market Tax Credits—not

technically an ecosystem service market but still a financial incentive from the public sector that provides financial incentives for enhancing the public good.

Conclusion

The bifurcation of forest management between endangered species protection and intensive forestry in the Pacific Northwest has exacerbated false choices about conservation and development. These two opposing pursuits ignore an approach that explicitly seeks financial returns from both timber production and the enhancement of public values, such as carbon storage, water quality, and scenic vistas. This approach produces high-quality timber and a vast array of nontimber forest products and services and plays to the competitive strengths of the Pacific Northwest, where trees grow old while remaining productive and residents value the "second paycheck" that environmental services and amenities provide. Ecological forest management augments species recovery efforts on federal forests by enhancing forest diversity and structure on private lands and is likely to create more resilient forests that are better positioned to survive the looming changes in global markets and climate.

LITERATURE CITED

Binkley, C., S. Beebe, D. New, and B. von Hagen. 2006. *An ecosystem-based forestry investment strategy for the coastal temperate rainforests of North America*. Portland, OR: Ecotrust.

Carey, A., B. Lippke, and J. Sessions. 1999. Intentional systems management: Managing forests for biodiversity. *Journal of Sustainable Forestry* 9(3/4):83–119.

Cederholm, C. J., D. H. Johnson, R. E. Bilby, L. G. Dominguez, A. M. Garrett, W. H. Graeber, E. L. Greda, et al. 2000. *Pacific salmon and wildlife–Ecological contexts, relationships, and implications for management*. Special Edition Technical Report, Prepared for D. H. Johnson and T. A. O'Neil (managing directors), Wildlife–habitat relationships in Oregon and Washington. Olympia: Washington Department of Fish and Wildlife.

Curtis, R. 1997. The role of extended rotations. In *Creating a forest for the 21st century*, edited by K. Kohm and J. Franklin. Washington, D.C.: Island Press.

Haynes, R. 2005. *Economic feasibility of longer management regimes in the Douglas-Fir region*. Research Note PNW-RN-547. Portland, OR: USDA Forest Service, Pacific Northwest Research Station.

Smithwick, E. A. H., M. E. Harmon, S. M. Remillard, S. A. Acker, and J. F. Franklin. 2002. Potential upper bounds of carbon stores in forests of the Pacific Northwest. *Ecological Applications* 12:1303–17.

von Hagen, B., and M. S. Burnett. 2006. Emerging markets for carbon stored by northwest forests. In *Forests, carbon, and climate change: A synthesis of science findings*. Salem: Oregon Forest Resources Institute.

PART VI

Synthesis

This book has been a study of old growth as both a natural and a social phenomenon. We argue that efforts to understand the old-growth issue (if focused on only one of these inherent aspects) will miss the full story. Old-growth studies are really at the intersection of the natural sciences and the social sciences; through this research, we may come to understand the full complexity and challenges of forest conservation. We do not argue, as do some post-modern critics of natural sciences, that scientific concepts of forests in general and old growth in particular are only social constructs. We maintain that natural sciences provide a powerful combination of tools and a structured approach to understanding the essential biophysical character and behavior of old forests and can serve as a foundation for policies and management practices to conserve them. In turn, the social sciences use their own tools and structured approaches to consider the coupled nature of human and natural systems—in this case, the changing relationships through time between humans and the forests they inhabit, utilize, visit, or worship. The strong and complicated connection between science and society has risen to the forefront in the old-growth debate. It is this interaction that we turn to in these final synthesis chapters.

In the first synthesis chapter, we (Duncan, Lach, and Spies) identify three emergent themes in the set of twenty-five chapters. These themes all involve the strong coupling between natural and human systems. The authors whose work we synthesize were selected to represent experts from a range of ecological and socio-economic sciences. While our selection process was guided by a first approximation of a thematic structure (e.g., ecological, social, economic, management), the final organization and themes we identified emerged only as we studied the chapters and discovered the connections and common themes among them. But we wanted more than just perspectives from ecological and social scientists—we also wanted some voices that represented stakeholder groups (e.g., environmentalists and timber industry). The voices and styles in these chapters were more diverse than in typical academic studies, but this diversity is as important and real to the future of the forests themselves as is their ecological diversity. The first chapter of Part VI concludes with a call for a new way of thinking about forests that arises from dynamics and variation in both ecological and social realms.

In the final chapter of Part VI—and of the book—we (Spies, Duncan, Johnson, Swanson, and Lach) take this integrated approach to thinking about old-growth forests a step further and propose some strategies for moving beyond the old-growth icon to a deeper foundation for forest conservation in the Pacific Northwest. We summarize some of the problems with the status quo and suggest how management and our concepts might be reinvented to deal with the new world of old growth in the region. This will require us to approach the problem at multiple scales, incorporate new economic thinking, and pay attention to how people learn and deal with change. None of our suggestions will be easy to implement, but even incremental improvements along these lines will put us in a better position to deal with the changes that will come.

Chapter 26

Old Growth in a New World: A Synthesis

SALLY L. DUNCAN, DENISE LACH, AND THOMAS A. SPIES

During the old-growth wars of the late twentieth century, a multitude of voices protested, planned, persuaded, and preached around this divisive issue. Although it seemed then that only two ideas prevailed—trees were for cutting or trees were for saving—the passing of time has allowed complexities and details to be teased out of the various perspectives, rendering considerably more nuanced ideas about old growth. Our body of knowledge about old-growth forests continues to expand, and our understanding of how we as individuals and as a society relate to them continues to evolve.

This book introduces the idea that one of the lasting products of the old-growth wars is the politically untouchable icon of big, old trees. The icon had the short-term effect of hardening the outlines of what many people conceive of as old-growth forests and creating a glare that made it difficult to see important complexities and subtleties of meaning central to the bigger problem of conserving forest biodiversity in a socially sustainable manner. It also had the effect of binding us to some principles of conservation that have arguably backed us into a tight corner when it comes to taking actions in the forest to restore and maintain biological diversity and human connections to the land. Although most of the authors in this book did not directly address the notion of old-growth forests as icon, all responded to these consequences of the "iconization" of old growth. Questions continue to swirl around old-growth forests. We wonder if these questions represent a much larger twenty-first-century struggle to understand the tangled and messy connection between ourselves and our environment.

Three overriding themes emerge from the twenty-four invited chapters:

- Old-growth forests achieved their status as icon through a convergence of scientific complexity and spiritual mystery.
- Unlike the static view of old-growth forests as "complete," these forests continually manifest both ecological and social change.
- Issues related to old-growth forests reflect social values and how they play out in a context of science, economics, and politics.

Our changing understanding of these themes, especially the increased understanding of environmental complexity, suggests a twenty-first-century renaissance in the way we might manage old-growth forests. This chapter will first synthesize the findings of the authors and then consider some of the ideas that were not mentioned.

Complexity Meets Mystery

Old-growth forests achieved their status as icons—objects of uncritical devotion—through the convergence of our increasing scientific understanding of complex systems with the sense of awe and mystery many already feel in the presence of big, old trees. Scientists increasingly view old growth as a manifestation of complex systems, one that challenges the way we traditionally practice and think about science and forestry. Instead of trying to understand all the "parts" of old-growth forests—species and vegetation types, for example—in the hope that the sum of those parts will tell us about the whole that is the forest, scientists are recognizing the importance of scale, time, and space as dynamic variables that affect how we perceive the forest. And, as in most complex systems, forests have emergent properties—structure, functions, and habitats—that are not predictable across scales because of the particular relationships among the parts, processes, and context. To scientists, old-growth forests represent the myriad intricacies and diversity of ecosystems.

Although old trees are a necessary component of diversity, they are only one part of the web that is the forest and are not sufficient by themselves. New scholarship reinforces the importance of each successional stage, not just old growth, in the life of the forest. The idea of abandoning the exclusive focus on old growth as a stage in forest development emerged strongly from several of our chapters. Indeed, in the view of many authors, simplifying forest types in any fashion serves to endanger complexity, and thus, by default, to endanger biodiversity, which appears to depend more on forest complexity than it does on forest age. And complexity, as it is now better understood, ought precisely to be the most cherished aspect of any forest not managed for exclusive industrial use. Multiage management and intentionally managing forests to contribute to complexity are complemented by recognition of the importance of the natural young forests that have been as dramatically reduced across the Pacific Northwest landscape as old growth—another example of the danger in simplifying forest types. These forests make crucial contributions to regional biodiversity that are as critical as those from old growth.

Each old-growth forest also has what we might call a *vintage*—a term borrowed from the wine industry that reflects not just a time when it was born or came of age but also a place. No two old-growth forests will ever be exactly the same because each developed through a unique environment and history. To complicate factors even more, we are also starting to see a change in the rates and patterns of change itself as humans speed up (e.g., rates of disturbance) or truncate (e.g., fire frequencies, invasive species) processes critical to forest development. Scientists are wondering which, if any, characteristics of old growth are invariable across place and time. Structural complexity or "messiness" may be the best distinction between old and young, managed and unmanaged forests that we can devise at this time.

The excitement generated as scientists increase understanding of complex forests is matched by a sense of wonder that many nonscientists feel when entering a grove of old-growth trees. To nonscientists, the complexity of old-growth forests can represent a refuge for the imagination in a materialistic world where everything is counted and accounted for. Even the language used to talk about old trees—*ancient, cathedral, old growth*—reflects a traditional, mythic understanding of the world that holds respect for elders and their experience-based knowledge. As Moore reminds us in her chapter,

> Old-growth forests are old. At least in the Pacific Northwest, they are tall. They are complex. They are unspoiled. They are quiet. They are beautiful. They are all of these at once. These are their sources of spiritual value.

We wonder what those trees, some standing since before Columbus crossed the Atlantic, have seen over the centuries. If they could speak, what would they tell us about our history and our future as users of the forest?

At the height of the old-growth wars, early scientific understanding of the complexity of forests converged with the spiritual and aesthetic understanding of the place that forests hold in our history, imagination, and culture to create an icon of old-growth trees as something special to be protected at all costs. If old-growth forests are effectively irreplaceable, as many believe, should we not protect them now and into the near future? And, until science could give us a better explanation of how forests actually worked, putting old-growth forests into reserves that protected them from human encroachment seemed a reasonable approach. Ultimately, however, the iconic status of old growth could not protect forests from ecological change or evolving social values.

Change Is Now and Forever

A strong component of old growth as icon is the notion of constancy—once an old-growth forest, always an old-growth forest—ignoring the realities of both ecological and social dynamics. Yet the importance of change was a substantial element of almost every chapter and surfaced as a key theme for the book. The dynamic nature of forests has emerged as one of the most compelling new perspectives of ongoing ecosystem research during the past three decades.

As noted by many authors, the primacy of the idea of a "climax" forest has run its course, and the concept is now frequently used as an illustration of how far we have traveled from earlier understanding of forests as staged and ordered. The static idea of a "climax" old-growth state was misleading in several respects and can be discerned behind the ecological and social consequences of policies such as "multiple use," continuous harvest, and even the more recently imposed old-growth reserves of the Northwest Forest Plan.

If we are convinced that old growth is a final "stage" in a linear sequence of forest development, we can comfortably lock up that specific group of trees for protection and move on. If, instead, we think that old-growth characteristics typically occur in a web of change in which the forest moves forward, backward, sideways, and inside out in response to disturbance and ecological development, we need to find new ways to manage the forest and also ways to understand change itself.

Several of the chapters point out that active management is needed to restore old-growth diversity in many cases; solving the issues of forest conservation and restoration cannot be addressed through isolation and preservation of a single stage of the forest. When we reframe the issue as encompassing multiple stages and types of forests, it is more apparent that forest conservation is something to which all landowners can contribute. Young, managed stands can be manipulated in ways that enhance the biodiversity of elements that are often characterized as old growth (e.g., species distributions, vegetation types).

Politics, economics, and cultural conditions, interacting with each other and the environment in various combinations, have also changed over the past thirty years, creating an equally strong dynamic affecting what we expect from forests. And, as noted by many authors, resistance to change has informed most environmental action, and the conflict over old growth is no exception. Ironically, this observation explains both the industry and the environmentalist positions: Industry supporters did not want to see their economic circumstances change in the drastic ways they anticipated, and environmental supporters did not want to see the timeless symbols of the region changed beyond recognition through continued clearcutting of old trees.

One outcome of the old-growth wars, courtesy of the Northwest Forest Plan, was to develop Adaptive Management Areas, where this wicked problem was to be taken on incrementally, in a learn-as-you-go mode. However, this experiment to manage continuous ecological and social change has not lived up to its promise. To date, AMAs have foundered on the risk-averse orientation of stakeholders, including many managers, environmentalists, and regulatory agency personnel. Challenges to the status quo, even ones that attempt to apply place-specific learning, are rebuffed through delay, denial, and outright hostility to new suggestions. This phenomenon is not unique to the forestry realm but rather one of the negative consequences of formalizing and institutionalizing practices that for the most part do help in getting work done efficiently and effectively.

Even as the old-growth wars heated up, technological changes were already playing a part in new approaches to forest management and were symptoms themselves of ongoing social change whose effects are just as far-reaching as those of large wildfires, regionwide insect outbreaks, or a new set of reserves in the affected ecosystems. The move by the end of the 1980s to an "agricultural model" of private forests (e.g., improved genetics, fertilization, density optimization) produced trees that could be processed by new and highly efficient mills that used second-growth timber grown on private lands to produce uniform logs for emerging markets. Thus, a complicated combination of private land management and mill ownership patterns, plus federal rules prohibiting export of federal logs, had created a thriving wood processing industry that depended on the larger older logs from federal land. When the federal timber supply dried up at the end of the old-growth wars, many mills without access to private timber were unable to retool quickly enough to compete for private timber. The Northwest economy as a whole grew strongly during this period, but as the older mills closed, some rural communities went into economic and social tailspins as they grappled with declining jobs, populations, and economic opportunities.

Social Values, Institutional Values, Whose Values?

The strong role of science in the old-growth debate gives the impression that this is a scientific problem, perhaps crossed with economics. Ultimately, the authors tell us, the old-growth wars were not about scientific advances or specific stands of trees. They were about values and the complicated humans who hold those values. In many ways, the social resonance of old-growth forests reflects the need humans have for experiences that take us beyond our immediate materialistic world. This moves us into the realm of spiritual values, which only amplifies the force of the old-growth icon in social debates.

It is not difficult to imagine that the future of old growth will remain dependent on social factors, not scientific or technological ones. Forests in general, and perhaps old growth in particular, have long been defining factors in the sense of place of the Pacific Northwest, for both rural and urban inhabitants of the region. And therein lies the rub. Forests are valued differently by people who earn their living through forests and their products, people who reside near them, and people who live in removed or even distant urban settings. As environmental values extended their reach into both urban and rural communities, many Americans were first learning about public forests through the lens of the old-growth icon.

The idea that people with different values could have a say in the fate of federal forests generated a great deal of anger among both rural and urban residents. Rural communities had become accustomed to living on the revenues provided by nearby national forests and resented those who wanted to "lock up" the forests for nonutilitarian purposes and what were often viewed as personal reasons. Urban environmentalists tended to characterize rural residents as exploiters of the common good and despoilers of the "last great places." These polarized positions contributed to and were modified by the growing general public expectation that federal lands could provide values other than timber. As a consequence of the public shift in preferences for noncommodity use of national forests (e.g., habitat protection, recreation, aesthetics), the stability rural communities had realized through reliable federal timber harvests disappeared and is proving difficult to replace in many locations.

New methods for valuation of old growth and other natural resources are starting to emerge to address peoples' different views of both commodity and noncommodity forest products and services. As our authors suggest, any revaluation should reflect the mystery and enchantment that many people feel in the presence of old trees if it is to reflect the full range of individual and social responses to old growth.

The old-growth turmoil of the last decades of the twentieth century is a symbol of large changes in our scientific and social understanding of the world. Considerable agreement exists among our authors that old growth as icon was solidified through changing ecological understanding of the complex forests that many viewed with awe approaching the religious. Changing technology as well as political alignments exacerbated the social change driving a reassessment of the value of forests for many who neither lived near nor were reliant on forests for revenues. The cross-cutting issues identified by these authors provide a thoughtful examination of the current state of our understanding of old-growth forests. Several issues, however, were either only briefly discussed or not mentioned at all and are worthy of consideration.

Emerging Issues

Among the core drivers of landscape change, increasing human population in the Pacific Northwest and the very nature of human behavior must surely be considered key to the likely future of most natural resource management. Whether growing cities and increasingly urbanized rural areas lead to a vast encroachment on resource lands, including forests of all kinds, or a more intense focus on the high value of remaining old-growth stands, the point is that old-growth forests will be affected either way. Population growth *will* have an impact on how and whether we manage these forests, and we would be foolhardy to imagine otherwise. Furthermore, we should not be ignoring the forces that affect public policy; individual and collective human decisions—typically acquisitive rather than conservative—tell us more about the future of old growth than any polls supporting or opposing their preservation. Changes in land use, changes in climate, changes in population, and changes in values all get writ large upon the landscape.

The old-growth debate was conducted in some quite novel ways and forums, testing new ideas and bringing into question the potential for managing future natural resource issues. The power of new knowledge communities—interest groups that formed around either the whole old-growth issue or local pieces of it—has been growing during the past three decades. A great many factors contributed to this. Social ones include changes in values, strengthening senses of place, shifts in financial resources available to nonprofits at the national level, and new skills in issue identification and litigation. Technological factors include the spread of the Internet and all its communication capacities, new capabilities such as geographic information systems, and data mining contributing to new knowledge. The rise of small activist/collaborative groups and their ability to network via the Internet to know they are not alone, to build community and momentum, and to coalesce to form tangible political power suggest that they are becoming a force—albeit little recognized—driving significant social change. They will certainly be a continuing part of the old-growth issue, and the fact that some of them do not necessarily last very long is simply part of their narrative. Their mere existence raises questions about current and future knowledge management, its meaning for democracy, and its role in any twenty-first century decision- and policymaking.

The challenge of managing forests across ownership boundaries also has a quiet but powerful role to play in how we think about conserving native forest diversity, including old growth. A recent change in tax law, for example, has shifted thousands of acres of former industrial timber lands into the portfolios of private equity firms and related entities. How will the management intentions of such entities affect the trajectory of forest development across the larger landscape? How long these lands stay forested is more frequently a decision based on return on investment than on ecology. Policies for mitigating climate change are turning attention to the value of forests as carbon sinks, an ecosystem service that can be bought and sold in markets around the world. These issues help us remember that, in addressing social change, we tend to think in terms of economics, not ecosystems. We are as yet far from integrating those two large arenas in any meaningful way and certainly not in how they operate together or separately in the future and the management of old-growth ecosystems.

These, then, are some of the issues that swirl around the old-growth debate as it now stands, contributing a volatile mixture of complexity and further uncertainty to a dialogue that continues to energize some sectors of the regional and national population every time it's mentioned. What the future holds is unclear, but we do know that the multitude and reach of events that have occurred since the environmental legislation of the 1970s has led us into a deeply changed era. Can this convergence of ideas and events be part of a larger renaissance in views about the environment and our increasingly complicated relationship with it?

Reconceiving the Forest: A Twenty-First-Century Renaissance?

The old-growth crisis fomented vigorous argument and changed many of the ways we understand forests. But the development of the old-growth icon almost guaranteed the narrowing and hardening of arguments, challenging our ability to come to any kind of agreement on the future. One example of such narrowing can be seen in demands to protect large swaths of the federal forest regardless of their age, structure, or health. When an icon becomes as powerful as old growth has, people can and will use it for all sorts of purposes. The thirty years since the beginning of the old-growth wars suggest that we could limp along without a policy fix for some time yet. But does our collective indecision reflect a far more troubling inability to think clearly about this and multiple future environmental tasks?

It does appear that we have many of the compass points required to define a renaissance of ideas about forest management: a fundamental upheaval in entrenched forest practices, a challenge to strong sets of existing scientific and social beliefs, a major shift in values both geographically and socially, new social and political communities, and a recognition that policy-making skills are not always up to the task. The old-growth wars may just have been a jolt that emerged at a time of multiple small revolutions, leaving us with the impression that they were the genesis of change, when in fact they were merely one symptom. And if indeed they were a symptom of fundamental change in how we think about environmental conflict and management, then it will be important to consider what new compass points we might follow for improving our orientation toward the future of managing ecosystems.

Complicating any future efforts to find solutions for complex environmental problems will be our ability to make decisions when complexity and uncertainty increasingly refuse to dissolve with time. Pursuit of knowledge, through the scientific method, as well as through less-formal or structured means, will continue to generate numbers, ideas, and questions. Indeed, the growing masses of data may bury us under their sheer volume. If we're thoughtful about it, however, they can help more rigorously outline the scope of the questions we must address.

This is clearly the challenge with how to manage, or even think about, old-growth forests. For wicked problems such as how to conserve old growth and at what level, decision making is not typically rational. The most likely solutions will probably be "clumsy" attempts to cobble together policies that are acceptable to a wide range of constituents. Our ability to grapple with complex issues is poorly developed, and the classic fall-back position of *symbolic politics*—more gesture than substance—has been shrewdly utilized in the old-growth management realm, a clear indication that it is too difficult to address directly.

But address it we must, for we are facing relentless social and environmental cycles that cannot be stared down. People will not abdicate their positions on old growth any time soon, and no matter how fervently some scientists may believe it, providing more scientific information will do little to change opinions. Furthermore, even supposing climate were a completely stable concept, the forests themselves will continue their cycles of change, established through millennia as the inevitable forces by which ecosystems evolve. If we add in whatever degree of climate change emerges in coming decades, a renaissance in how we think about forests will not only be likely, it will be essential. We need to think about forests as entities that change, *all* the time—at different rates, with no fixed ages or fixed structures. We need to think about social values that change, *all* the time, fluctuating around multiple and dynamic cultural, political, economic, and scientific ideas. We need to embrace the idea that science can provide us with many questions and many, although not all, answers. Science, however, cannot measure awe or mystery, and we are continuously reminded that it does not own all the measuring tools.

So we need not just to reconceive the forest as a dynamic and diverse system but to take our cue from what we have learned about old growth and reconceive forest management as a dynamic process that draws on multiple and often divergent knowledge, practices, and social expectations of natural resources. Changes in the way we think about both old growth and management have the potential to revitalize the role forests play as critical natural resources with the power to meet ecological, economic, and cultural needs of society. The challenge now is to move past the obsolete terms of the old-growth wars, prepared to embrace new and more demanding ways of thinking, and to explore untested approaches.

Chapter 27

Conserving Old Growth in a New World

THOMAS A. SPIES, SALLY L. DUNCAN,
K. NORMAN JOHNSON, FREDERICK J. SWANSON,
AND DENISE LACH

The most recent notion of old-growth forests captured the public imagination thirty years ago. An icon emerged, often as a romanticized vision of big, old trees, and environmentalists, policymakers, and scientists grabbed onto it to promote various competing agendas. But despite three decades of battles, conservation plans, and declarations of victory, there is no social consensus on old growth. It remains a mystery to many, a conundrum to managers and policymakers, and an unsure bet for those hoping to "preserve" it without change. Though the fires of old-growth conflict have died down, the issue still smolders, always threatening to flare up again.

The old-growth quandary will continue to dog us because it is so complex: there is little agreement on definitions and never will be, there is a great deal of geographic variation in old-growth forest structure and composition, the continuous changes within forests themselves foil attempts to manage exclusively for old growth, and no one knows how mounting human impacts will affect forests in the future. And most challenging of all are the fundamental differences in personal worldviews.

These problems make it difficult to get agreement, inventory the remaining areas of old forest, and sort out the "save it or cut it" arguments that have driven the old-growth wars in the past. We're coming to realize that the overwhelming challenge for land managers—public and private—

is not to maintain old-growth forests at any cost, but instead to protect and foster complex forest ecosystems in a world of unending social and environmental change. Policymakers who cannot stay on top of those waves of change may be plunged once again unprepared into future upheavals.

And the upheavals will come.

To even begin preparing for the future, we need to understand several difficult things:

- Society will never agree on a definition of old growth;
- The use of icons for complex ecological/social phenomena leads to one-size-fits-all short-term solutions and static thinking, and comes with ecological and social costs; and
- Current policies are layered and limiting, tattered and failing, and have many unintended consequences.

Our remaining old forests are both inspiring for the opportunity they provide for learning and humbling for revealing our inadequate understanding of the mysteries of forest ecosystems and how humans can best live with them.

Definitions

Given the diversity and dynamism we document in this book, we do not believe it will be possible to define old growth in a way that is agreeable to all parties. No one person or group "owns" the definition of old growth: it is a logical expression of the diversity of ways humans view forests and the diversity of the forests themselves. "The belief that nature is or can be measured and described before one decides what is important . . . is a dangerous illusion" (Norton, 1998). This belief might lead to defining old growth very simply— either by a single age (e.g., 200 years) or by a diameter of the tree (e.g., greater than twenty inches)—but such an approach does no justice to ecological and geographic diversity (fig. 27.1). Defining old growth and our goals for it will have to be a continuous process that engages managers, society, and scientists. Working definitions based on structure, composition, and processes are still needed by managers, but the idea of a final consensus is a fantasy. Provisional definitions should also recognize that the forest world is not black and white, that there are many shades of gray when it comes to old forest diversity, both in terms of the social landscape and the ecological landscape.

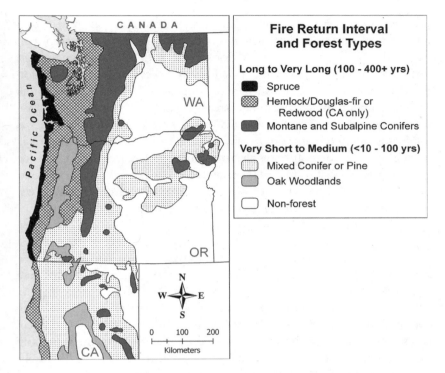

Fire Return Interval and Forest Types

Long to Very Long (100 - 400+ yrs)
- Spruce
- Hemlock/Douglas-fir or Redwood (CA only)
- Montane and Subalpine Conifers

Very Short to Medium (<10 - 100 yrs)
- Mixed Conifer or Pine
- Oak Woodlands
- Non-forest

FIGURE 27.1. Fire frequency and forest-type regions of Washington, Oregon, and Northern California. Old-growth forest structure, dynamics, and definitions would differ across these regions. The need for active management to reduce fuels and modify structure would be greatest in the mixed conifer, pine, and oak types.

Moving Past the Icon

We are challenged to come to terms with a number of annoying facts, not the least of which is the geographic variation in old growth and the importance of disturbance in its development. Old growth is not only seas of huge and ancient trees. We can certainly find examples of these in the region, and landscapes where the stumps bear witness to their existence in the past. If we keep traveling mindfully through our unlogged forests, however, for every acre of iconic old forest we find another acre where the big trees aren't so big, numerous, or old, and also patches where disease, wind, and fire have created openings and young forests. Further probing leads to whole watersheds untouched by modern humans, other forests that are young or middle-aged, and still others where no old trees can be found except perhaps as decayed fragments of wood from previous forests. And then the pine forests of central Oregon or the mixed-conifer forests of northern California show us other expressions where old trees are fugitives from fire, either mingling among patches of younger trees or hiding out in groups on moist, north-facing slopes. The aversion that many have to the death of trees at the hands of either nature (e.g., through fire or insects) or humans (e.g., through harvest) demonstrates that our innate preferences for life and stability impose a socially inflicted set of blinders. For example, the iconic view of old-growth forest as irreplaceable might drive an attempt to keep patchy fires, fuel reduction, or thinning out of all old-growth forest reserves.

Our fascination with old growth also might lead us to believe that it is the only stage of forest succession that is important to native forest diversity—it is not. Other biotic communities, such as open semipermanent meadows and young ecologically diverse forests, are also important and even rarer than old-growth forests in the Pacific Northwest region. When we put conservation dollars and efforts into protecting old-growth conifer forests, are we neglecting the needs of other threatened ecosystem types such as oak woodlands and lowland riparian forests, which are both threatened by development, or are we focusing too much on the public lands without considering the entire forest landscape?

Unintended Consequences of Current Approaches and Policies

Policies based on simple thinking and icons can lead to ironies and unintended consequences such as these: The forest industry—often seen as the problem—is essential to restoring and maintaining old growth; protecting spotted owl habitat in some fire-prone landscapes can increase the risk of high-severity fire in old growth; and protecting our older forests in the Pacific Northwest has just shifted U.S. sources of wood to natural older forests in Canada. Furthermore, our current federal policies were intended to resolve the controversy over how to protect the remaining old growth and provide some income and jobs for local communities. But they are not working that way. The recent policy may be a victory for old-growth conservation in the short run, but it generates mounting pressure (e.g., the lawsuit that forced the development of new plans that would cut more old growth on Bureau of Land Management lands) to produce more wood from federal forests, so it could backfire for old-growth conservation in the longer run. In short, our policies are a patchwork of Band-Aids with unintended consequences.

Strategies to protect forest diversity that rely heavily on listed species may be politically powerful but ecologically weak and unrealistic. Species approaches can give the impression that stability is the norm when in fact stability is not characteristic of natural populations any more than it is of natural ecosystems. The case of the spotted owl, whose populations are now threatened by both the barred owl and habitat loss, demonstrates that the use of single indicators or flagship species for whole ecosystems has severe limitations. Species approaches do still have an important role to play, and a species can be a useful barometer of ecological change, but if we rely too heavily on them we may miss other signs and other measures that are equally if not more important in our struggle to conserve our native forest diversity.

Finally, another type of icon needs to be addressed here. It is the icon of the perfect "answer" to the old-growth conservation question, wherein we are bedeviled by the expectation that neat technical solutions can be found to such complex ecological and social issues. This delusion further complicates the wild and messy social and ecological landscape that is old growth.

Opportunities and Uncertainties

Despite the logging and burning of more than twenty-five million acres of forest in the area in the past 100 years, the Pacific Northwest and northern California are relatively rich in centuries-old forests: at least three to eight million acres, depending on how you define old growth. This forest legacy of nature and the decisions of humans who have gone before us create values and opportunity. The values—both ecological and social—derive from the national and global rarity of old-growth forests and their distinctive biodiversity and appearance. The opportunity lies in *creatively* using knowledge of these forests to improve the sustainability of all forest management in our region.

But these learning opportunities are tempered by some significant conceptual puzzles and uncertainties. For example, forest time is a mystery. We can speak of the age of a Douglas-fir forest in western Washington as being 500 years—the time since the large fire that started the canopy cohort of this sun-loving species. This "500-year-old" forest is actually composed of organisms that range from less than one to over 500 years (most are much younger than 500 years), and organic matter that ranges from less than one to over 1,000 years in age. It is even more difficult to speak of the "age" of the drier forests, where the age of the "old" trees spans hundreds of years, reflecting numerous patchy fires. No one can say when the forest actually "began." For the human mind and a modern society founded on the concept of beginnings and endings and short-term thinking, forest time is not easily comprehended. Perhaps aboriginal peoples, like those of Australia, who believed in different forms of time—for example, dreamtime—could more easily fathom forest time.

How can we have the audacity to think that we can manage and conserve a complex natural process that requires thinking over centuries about many different things? Management often requires shortcuts and simplifications to make actions feasible or economical, but such efforts can be antithetical to ecological complexity. On the other hand, given the amount of change we have already brought to the forests and landscapes of this region, how can we assume that these forests will be just fine if we quietly walk away and leave them alone with benign intent? Many old trees in fire-prone landscapes have been lost to fire and insects because of the effects of past human actions—effects that often now require rectification. There is no right answer to the questions posed above. No experiment can be done that will run for 500 years and then tell us how we should have done it in 2008 or should do it in 2508 after centuries of environmental and social change.

Such uncertainty may suggest we be guided by the precautionary principle, which puts the burden of proof for environmental decisions on those who would take action. But what if no action—for example, *not* reducing the forest fuels we contributed to by suppressing fire—is itself a profound action?

Despite these enigmas and uncertainties we find hope in the millions of acres of unlogged older forest that remain. This land lets us hope that we have sufficient viable old-forest ecosystems with which to face growing human populations, fire, development, and climate change. Without these large tracts, the questions of viability and significant restoration would be moot; for in most settings, when modern human societies impose themselves on old forests, the results are inevitable. We believe our millions of old-growth acres in the Pacific Northwest can help us achieve a different outcome.

In this final chapter, we want to consider our management options, blinders off.

Reinventing Management

We propose that any future old-growth management strategies have four elements:

- Rethinking reserves
- Thinking across landscapes and ownerships
- Developing a new economics
- Learning from and adapting to change.

Rethinking Reserves

Reserves or areas dedicated to the protection and restoration of old forest will always be part of an old-growth conservation strategy. But we need many kinds of reserves. For example, strict reserves are needed where natural processes can operate without the direct influence of humans. Such places can provide for old growth in forests where ecological processes are relatively unaltered by human activity, serve as controls for human-dominated forests, spread risk associated with the uncertainty of our management plans, and provide places of mystery and spiritual connection. Such reserves are most appropriate in wetter forest types where fire is infrequent (fig. 27.1). But many old-growth reserves will still be directly influenced by external processes such as wildfire, insects, invasive species, and climate change. A newer type of reserve, the ecological reserve or conservation area, is also needed, in which the particular ecological goals are identified and active management is used carefully to help achieve those goals. Active management in these reserves could include several tactics: thinning and mechanical removal of smaller diameter trees and eventually prescribed fire to reduce fuel levels and the risk of high-severity fire, planting or reintroduction of species that have been lost because of past human activities, and removal of invasive species either mechanically or even through chemical means.

Generally, in the case of ecological reserves, the larger and the more numerous the better. This assumption guided the development of the Northwest Forest Plan, where large (100,000+ acre) reserves were established in a network that was thought to have enough of a safety factor to allow old forests to persist. At first look, during the first ten years of the plan this assumption seems to have been borne out. The plan assumed some losses to wildfire across the network of reserves. A closer look at what actually happened, however, reveals a disturbing trend. Within the fire-prone forest types, the losses of old forest during the first ten years were quite high: ten to fourteen percent in some fire-prone provinces. If these rates were to continue, in fifty years little old growth would be left in these areas.

In general, our current approach to reserves in fire-prone landscapes does not appear to be up to the job of conserving and restoring old growth. Fire, insects, and disease connect these landscapes in ways that make artificial delineation of them into "reserves" and "nonreserves," a recipe for failure. Indeed, the entire public forest landscape is needed to maintain and produce fire-resilient older forests. Instead of setting conservation goals in terms of acres of reserves, we need goals in terms of landscape distributions of forest conditions that provide for a diversity of successional stages that have different habitat potentials and degrees of resistance to fire, insects, and disease. We will probably never reach those distributions—nature will be a comanager with a mind of her own—but these goals can be a kind of receding horizon toward which forests move. If the goal is the conservation of old forests, old trees, and species associated with them, then active intervention is needed within many (but not all) fire-prone old-growth forest reserves. The Northwest Forest Plan for the federal lands recognized this, but it has not been implemented; the rate and extent of fuel treatments needed in fire-prone landscapes is not keeping up with the need. Furthermore, the exact amounts and patterns of these activities cannot be generalized. A one-size-fits-all approach does not work, and both local landscape and site knowledge are essential.

An additional motivation for rethinking the reserves stems from the fact that current reserves are centered on the remnant old growth—the remains of a more than a century of logging. Consequently, the reserves are located based on management history and not necessarily on the best pattern to conserve old forest diversity in the long run. Under the current design, high productivity sites and low elevation sites are underrepresented in the reserve network.

Active management in and between reserves for ecological goals is not just for the drier fire-prone systems. Plantations that were originally intended for intensive timber management but now fall in ecological reserves may need restoration thinning. These plantations can be structurally diversified by using variable density (not uniform) thinning, which can simultaneously speed the development of some species associated with older forests. Thinning would also provide some valuable revenue for local communities. Producing wood from federal lands through thinning, which has widespread support, has been the most successful example of integration of biodiversity and commodity goals in the Northwest Forest Plan.

Thinking Across Landscapes and Ownerships

It is one thing to put lines on a map and say here we will produce biodiversity (i.e., *reserves*) and here we will produce timber (e.g., *matrix* also known as *general forest*). It is another to actually provide all of our forest values on the ground. What happens to this simplistic equation when we add disturbance and succession and organisms that do not happen to occur in the reserved areas? Adding these terms to our calculus requires that we also think more broadly, about whole landscapes. For example, the current approach to implementing the Northwest Forest Plan on federal lands is not sustainable over the long run. Much of the expected timber is now actually coming from small trees in plantations that are checkered across reserves. The volume that was expected from forests outside the reserves was slated to come from cutting older, larger trees. This has not happened. Once these plantations from the 1950s and 1970s reach eighty years, they will no longer be eligible for restoration thinning under current policies. The rate of thinning of these plantations on federal lands has not been high enough to affect all plantations before they reach eighty years. Consequently, restoration here will fall short of its potential.

Although increasing the budgets for this type of thinning on federal lands and using structure-based rules instead of age-based rules could alleviate this problem, it will not alleviate the longer term problem of the highly variable social acceptability of timber production on federal lands. Our best hope for conserving old-growth values and providing other forest goods and services may be through landscape-scale thinking, which implies thinking across all forest lands—not just reserves, not just public lands—and thinking in longer time frames—not just a few decades. In a different context, Dwight D. Eisenhower said, "If a problem can't be solved, enlarge it."

Much of the old-growth discussion typically revolves around federal lands. But the challenge of conserving native forest diversity cannot be met with either old growth or federal lands alone. Research has shown that many species associated with older forests do not stay within a single ownership and can find habitat in other forests that are structurally or compositionally diverse. It appears that much of our native biodiversity can be retained in landscapes that also contain forests managed for timber production, provided those management schemes pay attention to ecological diversity. We have learned that forests develop and occur with incredible variety, the function of which we do not fully understand. We should be careful in assuming that our simple classifications of this richness (e.g., "old growth" and "not old growth") are the best foundation for conserving this diversity. For many reasons, not all forest lands, public or private, will be dedicated to producing classic old growth. However, these far-less-iconic forests can still make important contributions to conserving forest biodiversity.

To a significant degree, the fallout from the old-growth wars of the 1990s has turned our view of the role of the forest industry on its head. Restoration activities in both fire-prone and wetter forests need loggers, logging systems, and mills to pay for and process the wood. Large-diameter wood, once a highly sought and valued commodity from old-growth forests, is no longer so highly valued. The decline in supply of big trees has not increased demand and price but has instead contributed to a trend in which smaller diameter logs from plantations are now sometimes more highly valued than the big trees because they are used in manufacturing large wood beams. With modernization and the decline of federal timber, the number of lumber mills has also declined, especially in the drier, eastern parts of the Pacific Northwest region. The loss of this mill and woods-worker capacity jeopardizes the ability to conduct restoration thinning and fuel reduction in these areas. In a continuing spiral, as the distances between the forest and the mills increase, the transportation costs go up and the profit margins go down, so that removing small-diameter wood (the kind that constitutes most of the fuel buildup) becomes even less economically viable (Lettman, 2007). More mill closures would create a vicious cycle.

In our approach to fire-prone forests, we need to focus on managing stands in new ways and at landscape scales. For example, rather than expecting to cut larger fire-resistant old-growth pines and Douglas-firs from the overstory, we should focus on managing the understories—the smaller diameter trees and the shorter lived shade-tolerant species (e.g., grand fir). Such approaches can reduce fuels, regenerate the old pines and Douglas-firs that we lost through fire and logging, and create the denser forests favored by the spotted owl in places across the landscape where it is most likely to survive fire. Much of this activity will not be economically viable in terms of the old standards for wood products, but some larger economically viable non-old-growth trees can be found. In this context, wood products become a byproduct of creating more fire-resilient forests and restoring old-growth conditions. We have reached a deeply ironic point in the history of the Northwest forest industry: Logging of old growth has nearly stopped, but losses to fire have increased in recent decades, and a robust forest industry is now needed to save the old pines, Douglas-firs, and western larches that were the former dominants of the drier old forests.

Developing a New Economics

Reconceiving our approaches to old-growth management relies on the third leg of our strategy: a new economics. The recognition of the new biodiversity and spiritual values related to old-growth and natural forests is not matched by a similar change in the accounting of those values in our economic systems: biodiversity, clean water, recreation, and spiritual values. Economists call these "existence values," for they cannot be translated easily into goods. That the existence values of old-growth forests are high is abundantly clear in the protests, headlines, and policy changes of the past twenty-five years. Yet many believe public-lands budgets to restore and protect these forests remain inadequate. More focused efforts to demonstrate the magnitude of ecosystem values to the U.S. government might generate more support for the actions needed to conserve and restore old forests. Ecosystem services and values can be captured in markets such as the emerging carbon-trading markets in Europe and the United States, although they are still being developed, and their future and effectiveness is uncertain.

One challenge is to make the carbon that is traded more tangible and visible to buyers — in other words, to make it more marketable. Given that there are few more-attractive forms of storage of carbon than an old-growth Douglas-fir forest, might the old-growth icon have another task to add to its job description?

The economic barriers to improving the biodiversity side of the ledger in a sustainable way on private lands are also significant. For example, growing trees on longer rotations, although producing more wood in the long run and storing more carbon, can be a significant cost to a landowner compared with short rotations. One way of mitigating this economic hit is through supplementing the income of landowners who undertake more costly ecological management approaches through the emerging ecosystem services markets.

The degree to which this new approach catches on is yet to be determined, but such efforts could fill an important niche and be promoted as scientific understanding and social forces align once again. We may witness a sea change as political currents related to fire, climate change, economics, and conservation merge. These currents include new social recognition of the need for fuel reduction in fire-prone landscapes; the need to reduce carbon dioxide emissions from the earth, including those created by large wildfires; the need to store carbon in forests—something old-growth forests and forests grown on longer rotations do very well—and the need to consider alternative sources of biomass such as wood products, which could become a less environmentally and socially disruptive source of biofuels than corn.

Learning from and Adapting to Change

Social and scientific views as well as the behavior of ecosystems are in constant flux. We cannot predict the exact nature, place, or timing of these changes, only that they *will* occur. Our experience in dealing with ecosystem management objectives is less than twenty years old—about one-twentieth of the lifetime of a typical old-growth tree. Adaptive management and joint learning thus form the final leg of our strategy.

Adaptive management was a cornerstone of the Northwest Forest Plan, but it has not developed as many would have hoped, even though monitoring efforts have produced much valuable and surprising information. For example, it was monitoring that showed that, despite an increase in the area of older forest, populations of northern spotted owls have still declined. But the process of monitoring biodiversity is new and can be expensive to agencies with budgets previously tied to timber production. It remains crucial, though, to maintain and even bolster monitoring efforts, basing them on clear measures of environmental indicators and expected trends in those measures. More needs to be done, particularly in terms of monitoring or validating the assumptions that underlie any conservation or restoration plan. Setting benchmarks, assessing and reassessing indicators, and evaluating progress toward objectives will provide information that will continue to inform and may routinely surprise us. At the very least, it will improve our ability to make midcourse corrections before unexpected trends can become policy fiascos.

Adaptive management and monitoring will not succeed, however, without overcoming significant social barriers. Most people and institutions prefer the status quo and resist learning. Developing trust among stakeholders is critical to moving adaptive management from the lips to the landscape; for without joint learning about complex ecosystem management problems, the chance of progress in dealing with these forest conundrums remains depressingly low. We will need new and trusted forms of social engagement if we are to achieve social license for changing management approaches. One way this can happen is through manager–public-scientist partnerships that focus on real places and implement alternative actions on the ground and evaluate them over time, including through public discussion.

There are several examples in the region, including the Central Cascades Adaptive Management Partnership on the Willamette National Forest, where some alternative approaches to meeting the goals of the Northwest Forest Plan are being carefully tested by collaboration among federal and university researchers, U.S. Forest Service managers, and highly informed members of the public. One experiment has attempted to use natural disturbance regimes as a guide for improving the effectiveness of maintaining native forest biodiversity at landscape levels and producing modest levels of wood to bolster economies in rural communities. Another example is the Lakeview Stewardship Group in central Oregon, where stakeholders in the debates about public forests are working together to find common ground in the management of fire-prone forests.

The social environment for forest management is a deeply fragmented landscape of individuals and interest groups. The only binding force for such disparate interests and approaches to learning appears so far to be a sense of place, where a commonly held vision of the future has the power to overcome risk aversion and keep long-held conflicts in perspective. The questions remains, how to achieve this on lands owned by all Americans and just a few Americans?

Old-Growth Futures

What might the future hold for old growth and our relationship to it in our changing world? We don't know the answer, but we can imagine some alterative possible, but not mutually exclusive, futures:

- After the past few decades of old growth as a *cause célèbre*, it may fade into the background. The present conservation strategies may be considered adequate and the topic deserving of little attention as other, more-pressing forest issues emerge.
- Old growth may continue to serve as an important source of scientific understanding and model of natural systems used in development of fresh approaches to forest conservation and the management of production forests.
- But, if the coming decades are at all like the past few decades, old growth will play an emotive role in society. Will the loss of big, old trees from environmental change and natural disturbances instead of from chain saws maintain our concern about human impact on this heritage and the environment in general?

Final Thoughts

What we need to do about old growth today must be done boldly but also thoughtfully. Each of the four related conservation strategies above calls for a continual process of learning, recognition of risk and uncertainty, and awareness that the complexity that scientists and managers see is also the mystery that enriches us. The government won't do this for us, although it might help along the way. Private corporations won't shower money on the problem, although they might be intrigued enough to talk. Granting agencies won't suddenly free up research budgets, although plenty of researchers have data and ideas already that they'd love to see getting used. And the "attentive" public? Some will stay attentive, and plenty of these people know way more than they're given credit for. No one can do this alone: We are all partners in this uncertain venture.

Although the sciences of ecology, conservation biology, and forestry have given us new understanding and tools to obtain many things from forests, we should be careful to avoid technological hubris that leads us to think that science and management alone show us the way to relate to these forests. And, in case we might think that the spiritual valuing of ancient forests is a discovery of Thoreau in the nineteenth century or the environmental movement of the 1970s, Lee (chapter 8) demonstrates the deep roots of the spiritual valuing of old forests when he quotes Seneca (Van der Leeuw, 1963), a Roman philosopher from 2,000 years ago, who wrote of the "presence of deity" in a "grove of ancient trees." It has been our hope, from the first conception of this book, that we might hold up the old-growth icon to turn it about in the light of our shared experience and reveal its many facets so that our readers might also see the compelling, complicated, alarming, encouraging, and challenging array of choices for the future.

LITERATURE CITED

Lettman, G. 2007. Personal communication. Salem: Oregon Department of Forestry.

Norton, B. G. 1998. Improving ecological communication: The role of ecologists in environmental policy formation. *Ecological Applications* 8(2):350–64.

Van der Leeuw, G. 1963. *Religion in essence and manifestation,* vol. 2. New York: Harper & Row.

CONTRIBUTORS

PETER A. BISSON is a research fisheries biologist and team leader with the U.S. Department of Agriculture (USDA) Forest Service's Aquatic and Land Interactions Program, Pacific Northwest Research Station in Olympia, Washington. He specializes in watershed management and native fish conservation.

RICK BROWN is a senior resource specialist with the Northwest office of Defenders of Wildlife, working primarily on conservation of forest biodiversity. He has particular interests in issues of forest management in fire-prone forests, and the implications of climate change for forest conservation. Defenders of Wildlife is dedicated to the protection of all native animals and plants in their natural communities.

ANDREW B. CAREY is an emeritus scientist with the USDA Forest Service, Pacific Northwest Research Station, Olympia, Washington, and owns and operates Serendipity Consulting in Ashford, Washington. He specializes in managing forests and landscapes for multiple values.

SALLY L. DUNCAN is a policy research program manager with the Institute for Natural Resources Directors Office, Oregon State University, in Corvallis. She specializes in the relationships between humans and their natural resources, especially the intertwined roles of science, communications, and social acceptance of resource management.

ERIC D. FORSMAN is a research wildlife biologist with the USDA Forest Service's Ecosystem Processes Program, Pacific Northwest Research Station in Corvallis, Oregon. His research interests include the demography of the northern spotted owl and the ecology of the red tree vole.

JERRY F. FRANKLIN is a professor of ecosystem analysis in the College of Forest Resources, University of Washington, in Seattle. He also is the director of the Wind River Canopy Crane Research Facility located in Carson, Washington. His research interests include stand development, ecosystem structure and function, and ecological forestry.

RICHARD W. HAYNES is a recently retired forest economist and program manager of the Human and Natural Resources Interactions program, USDA Forest Service, Pacific Northwest Research Station in Portland, Oregon. His specializations include U.S. forest sector models, the timber markets, and national forest timber demand.

K. NORMAN JOHNSON is Distinguished Professor in the Department of Forest Resources, College of Forestry, Oregon State University, Corvallis, Oregon. He specializes in forest planning and policy.

ANDY KERR is a senior counselor to Oregon Wild (formerly known as Natural Resources Council). He also runs the Larch Company, a nonmembership for-profit conservation organization. He lives in the recovered mill town of Ashland, Oregon.

DENISE LACH is an associate professor in the Department of Sociology at Oregon State University, Corvallis. She specializes in natural resource sociology and organizational sociology.

ROBERT G. LEE is a retired professor of sociology of natural resources in the College of Forest Resources at the University of Washington in Seattle. His research interests are human communities and development and change of forestry institutions.

JOHN LOOMIS is a professor in the Department of Agricultural and Resource Economics at Colorado State University in Fort Collins. He specializes in environmental and natural resource economics and public lands management. Some of the ideas in his paper arose from discussion among workshop participants. The concept of authenticity arose from discussions with Dr. Sandra Gudmundsen, professor emeritus at Metropolitan State College of Denver.

ROSS MICKEY is the western Oregon manager for the American Forest Resource Council in Portland. The group's mission includes creating a favorable operating environment for the forest products industry and ensuring a reliable timber supply from public and private lands.

KATHLEEN DEAN MOORE is Distinguished Professor of philosophy at Oregon State University in Corvallis and the founding director of the Spring Creek Project for Ideas, Nature, and the Written Word. Her current work is in the areas of environmental ethics and philosophy and nature.

BARRY R. NOON is a professor in the Department of Fishery and Wildlife Biology at Colorado State University in Fort Collins. His research interests include landscape ecology, population dynamics, conservation biology, and endangered species management.

CHADWICK DEARING OLIVER is the Pinchot Professor of Forestry and Environmental Studies and director of the Global Institute of Sustainable Forestry, School of Forestry and Environmental Studies at Yale University in New Haven, Connecticut. He is interested in how ecological knowledge of forests and silviculture can help resolve scientific, technical, and management issues at the landscape and policy levels.

JIM PROCTOR is a professor and director of the Environmental Studies Program with Lewis and Clark College in Portland, Oregon. He is interested in the relationships among nature, science, religion, and culture. He is grateful to the National Science Foundation (research grant BCS-0082009) for its support, plus that of two University of California, Santa Barbara, graduate students, Evan Berry and Tricia Mein, who worked alongside him in the survey (2002 survey cited in chapter 9).

GORDON H. REEVES is a research fish biologist with the Aquatic and Land Interactions Program, USDA Forest Service, Pacific Northwest Research Station, in Corvallis, Oregon. His research interests include aquatic ecosystems, landscape ecology, and salmonid conservation.

HAL SALWASSER is dean of the College of Forestry, director of the Oregon Forest Research Laboratory, and professor of forest resources and forest science at Oregon State University, Corvallis. He has been regional forester and research station director with the USDA Forest Service.

HOWARD SOHN is chairman of the board of Lone Rock Timber Company in Roseburg, Oregon. He is past chair of the Oregon Board of Forestry and serves as a director on the Oregon State Board of Higher Education.

THOMAS A. SPIES is a research forester and team leader with the Ecosystem Process Program, USDA Forest Service, Pacific Northwest Research Station, in Corvallis, Oregon. His research interests include forest dynamics, landscape ecology, and integrated approaches to assessing the ecological consequences of forest management.

GEORGE H. STANKEY is a retired research social scientist with the Human and Natural Resources Interactions Program, USDA Forest Service, Pacific Northwest Research Station, Corvallis, Oregon. His research interests include adaptive management and recreation planning.

BRENT S.STEEL is a professor in the Department of Political Science and director of the Master in Public Policy Program at Oregon State University, Corvallis. His research interests include environmental politics and policy.

FREDERICK J. SWANSON is a research geologist with the USDA Forest Service, Pacific Northwest Research Station, Corvallis, Oregon. His research interests include the role of large disturbance in ecosystems, and work at the interfaces among the sciences, humanities, and land management.

JOHN TAPPEINER is professor emeritus in the Department of Forest Resources at Oregon State University, Corvallis. His research interests include ecology and management of shrubs and hardwoods, silviculture, and development of old-growth forests.

JACK WARD THOMAS is professor emeritus in the Department of Forest Management, College of Forestry and Conservation, University of Montana, Missoula. He has been chief of the USDA Forest Service. His research interests include wildlife conservation and ecosystem management.

BETTINA VON HAGEN is vice president of Ecotrust's forestry program, Jean Vollum Natural Capital Center, Portland, Oregon. The mission of Ecotrust is to build a rain forest bioregion where economic, ecological, and social conditions are improving, where a "conservation economy" is emerging.

JULIA M. WONDOLLECK is an associate professor in the School of Natural Resources and Environment, University of Michigan, Ann Arbor. Her research interests include collaborative dimensions of ecosystem management, environmental dispute resolution, and public lands policy and administration.

INDEX

Note: Italicized page numbers indicate boxes, tables, figures, or photographs.